C000040140

Credits

Author

Vijay Parthasarathy

Reviewers

Robert E. McFrazier

Brian O'Neill

Ricardo J. S. Teixeira

Acquisition Editors

James Jones

Grant Mizen

Commissioning Editor

Manasi Pandire

Technical Editors

Gaurav Thingalaya

Dennis John

Project Coordinator

Suraj Bist

Proofreader

Paul Hindle

Copy Editors

Alisha Aranha

Dipti Kapadia

Sayanee Mukherjee

Aditya Nair

Kirti Pai

Shambhavi Pai

Lavina Pereira

Indexers

Rekha Nair

Tejal R. Soni

Graphics

Ronak Dhruv

Abhinash Sahu

Production Coordinator

Manu Joseph

Cover Work

Manu Joseph

About the Author

Vijay Parthasarathy is an Apache Cassandra Committer who has helped multiple companies use Cassandra successfully; most notably, he was instrumental in Netflix's move into Cassandra. Vijay has multiple years of experience in software engineering and managing large project teams. He has also successfully architected, designed, and developed multiple large-scale distributed computing systems, distributed databases, and highly concurrent systems.

I want to thank my wife, Narmadhadevi, and the rest of my family, who supported and encouraged me in spite of all the time it took me away from them.

Learning Cassandra for Administrators

Optimize high-scale data by tuning and troubleshooting using Cassandra

Vijay Parthasarathy

PUBLISHING

BIRMINGHAM - MUMBAI

Learning Cassandra for Administrators

Copyright © 2013 Packt Publishing

All rights reserved. No part of this book may be reproduced, stored in a retrieval system, or transmitted in any form or by any means, without the prior written permission of the publisher, except in the case of brief quotations embedded in critical articles or reviews.

Every effort has been made in the preparation of this book to ensure the accuracy of the information presented. However, the information contained in this book is sold without warranty, either express or implied. Neither the author, nor Packt Publishing, and its dealers and distributors will be held liable for any damages caused or alleged to be caused directly or indirectly by this book.

Packt Publishing has endeavored to provide trademark information about all of the companies and products mentioned in this book by the appropriate use of capitals. However, Packt Publishing cannot guarantee the accuracy of this information.

First published: November 2013

Production Reference: 1181113

Published by Packt Publishing Ltd.
Livery Place
35 Livery Street
Birmingham B3 2PB, UK.

ISBN 978-1-78216-817-1

www.packtpub.com

Cover Image by Faiz Fattohi (faizfattohi@gmail.com)

About the Reviewers

Robert E. McFrazier is an open source developer, manager, trainer, and architect. Having started with web development, he was able to progress in his career in multiple roles including developer, trainer, build/release engineer, architect, and manager. He has been working primarily in PHP web development, but he also has experience in Java development, AWS cloud, message queues, Hadoop, Cassandra, and creating high-volume SOAP/REST API services.

Robert has worked for many software companies including Nordstrom.com, InfoSpace, Clear, RealNetworks, and Arise Virtual Solutions.

> I'd like to thank my wife, Toni, and my son, Robbie, for being the inspiration to continue learning and growing.

Brian O'Neill likes to describe himself as a father, husband, hacker, hiker, kayaker, fisherman, Big Data believer, innovator, and a distributed computing dreamer.

Brian has been a technology leader for over 15 years. He has experience as an architect in a wide variety of settings, from startups to Fortune 500 companies. He believes in open source and contributes to numerous projects. He leads projects that extend Cassandra and integrate the database with full-text indexing engines, distributed processing frameworks, and analytics engines. He won the InfoWorld's Technology Leadership award in 2013. He has authored the Dzone reference card on Cassandra and was nominated as a DataStax Cassandra MVP in 2012.

In the past, Brian has contributed to expert groups within the Java Community Process (JCP), and he has patents in artificial intelligence and context-based discovery. Brian is proud to hold a B.S. in C.S. from Brown University.

Presently, Brian is the Chief Architect at Health Market Science (HMS), where he heads the development of their Big Data platform focused on data management and analysis for the Healthcare space. The platform is powered by Storm and Cassandra and delivers real-time data management and analytics as a service.

Brian is currently co-authoring a book scheduled to be published in December 2013, *Storm Blueprints for Distributed Computing, Packt Publishing*.

To my wife Lisa, it is in your sail's angelic wake that I find balance

You lift me up and keep me from capsizing. I love you

And to the silverback, for keeping my feet firmly on the ground

While the greyback let me dream among the clouds

To my sons, you are my inspiration, the wind at my back

The stars wait for you.

Without all of you, this ink would never have met this page.

Ricardo J. S. Teixeira is a Portuguese computer engineer.

He currently lives with his family in Vila do Conde, a quiet city by the Atlantic ocean. He is interested in how Big Data technologies can be used to solve common business problems.

Ricardo reviewed this book while he was writing a thesis for his Master's degree on leveraging Big Data technologies (such as Storm and Cassandra) to acquire near real-time metrics for a large retail store chain.

I would like to thank my parents, my brother, Milai, and Áurea for all the times they waited patiently while I was hunched over my laptop.

www.PacktPub.com

Support files, eBooks, discount offers and more

You might want to visit www.PacktPub.com for support files and downloads related to your book.

Did you know that Packt offers eBook versions of every book published, with PDF and ePub files available? You can upgrade to the eBook version at www.PacktPub. com and as a print book customer, you are entitled to a discount on the eBook copy. Get in touch with us at service@packtpub.com for more details.

At www.PacktPub.com, you can also read a collection of free technical articles, sign up for a range of free newsletters and receive exclusive discounts and offers on Packt books and eBooks.

http://PacktLib.PacktPub.com

Do you need instant solutions to your IT questions? PacktLib is Packt's online digital book library. Here, you can access, read and search across Packt's entire library of books.

Why Subscribe?

- Fully searchable across every book published by Packt
- Copy and paste, print and bookmark content
- On demand and accessible via web browser

Free Access for Packt account holders

If you have an account with Packt at www.PacktPub.com, you can use this to access PacktLib today and view nine entirely free books. Simply use your login credentials for immediate access.

Table of Contents

Preface   **1**

Chapter 1: Basic Concepts and Architecture   **5**

  CAP theorem   **5**

  BigTable / Log-structured data model   **6**

    Column families   6

    Keyspace   7

    Sorted String Table (SSTable)   7

    Memtable   7

    Compaction   8

  Partitioning and replication Dynamo style   **8**

    Gossip protocol   8

    Distributed hash table   9

    Eventual consistency   10

  Summary   **10**

Chapter 2: Installing Cassandra   **11**

  Memory, CPU, and network requirements   **11**

  Cassandra in-memory data structures   **12**

    Index summary   12

    Bloom filter   12

    Compression metadata   12

    SSDs versus spinning disks   13

    Key cache   13

    Row cache   14

  Downloading/choosing binaries to install   **14**

    Configuring cassandra-env.sh   14

Configuring Cassandra.yaml   15
cluster_name   15
seed_provider   15
Partitioner   15
auto_bootstrap   15
broadcast_address   16
commitlog_directory   16
data_file_directories   16
disk_failure_policy   16
initial_token   17
listen_address/rpc_address   17
Ports   17
endpoint_snitch   18
commitlog_sync   18
commitlog_segment_size_in_mb   18
commitlog_total_space_in_mb   18
Key cache and row cache saved to disk   19
compaction_preheat_key_cache   19
row_cache_provider   19
column_index_size_in_kb   20
compaction_throughput_mb_per_sec   20
in_memory_compaction_limit_in_mb   21
concurrent_compactors   21
populate_io_cache_on_flush   21
concurrent_reads   21
concurrent_writes   22
flush_largest_memtables_at   22
index_interval   22
memtable_total_space_in_mb   22
memtable_flush_queue_size   23
memtable_flush_writers   23
stream_throughput_outbound_megabits_per_sec   23
request_scheduler   23
request_scheduler_options   23
rpc_keepalive   24
rpc_server_type   24
thrift_framed_transport_size_in_mb   24
rpc_max_threads   24
rpc_min_threads   25
Timeouts   25

Dynamic snitch   26
Backup configurations   27
incremental_backups   28
auto_snapshot   28
Cassandra on EC2 instance   28
Snitch   29
Create a keyspace   30
Creating a column family   31

GC grace period	32
Compaction	33
Minimum and maximum compaction threshold	33
Secondary indexes	34
Composite primary key type	34
Options	36
read_repair_chance and dclocal_read_repair_chance	36
Summary	**36**
Chapter 3: Inserting Data and Manipulating Data	**37**
Querying data	**37**
USE	37
CREATE	38
ALTER	38
DESCRIBE	39
SELECT	39
Tracing	**40**
Data modeling	**41**
Types of columns	42
Common Cassandra data models	43
Denormalization	43
Creating a counter column family	46
Tweet data structure	46
Secondary index examples	46
Summary	**48**
Chapter 4: Administration and Large Deployments	**49**
Manual repair	**50**
Bootstrapping	**52**
Vnodes	53
Node tool commands	54
Cfhistograms	54
Cleanup	54
Decommission	54
Drain	54
Monitoring tools	**56**
DataStax OpsCenter	56
Basic JMX monitoring	57
Summary	**59**
Chapter 5: Performance Tuning	**61**
vmstat	**62**
iostat	**62**
dstat	**63**
Garbage collection	**64**

Enabling GC logging   65
Understanding GCLogs   65
   Stop-the-world GC   66
   The jstat tool   66
   The jmap tool   68
The write surveillance mode   68
Tuning memtables   **68**
memtable_flush_writers   69
Compaction tuning   **69**
SizeTieredCompactionStrategy   70
LeveledCompactionStrategy   70
Compression   **71**
NodeTool   71
compactionstats   72
netstats   72
tpstats   72
Cassandra's caches   73
   Filesystem caches   75
Separate drive for commit logs   75
Tuning the kernel for Cassandra   75
noop scheduler   76
NUMA   76
Other tuning parameters   77
Dynamic snitch   77
Configuring a Cassandra multiregion cluster   78
Summary   **80**
Chapter 6: Analytics   **81**
Hadoop integration   **81**
Configuring Hadoop with Cassandra   81
   Virtual datacenter   82
Acunu Analytics   85
Reading data directly from Cassandra   85
Analytics on backups   85
   File streaming   86
Summary   **88**
Chapter 7: Security and Troubleshooting   **89**
Encryption   **89**
Creating a keystore   90
Creating a truststore   90

Transparent data encryption   91
    Keyspace authentication (simple authenticator)   91
    JMX authentication   92
Audit   **93**
Things to look out for   **93**
Summary   **94**
Index   **97**

Preface

The Apache Cassandra database is the right choice when you need scalability in terms of millions to billions of rows with the highest availability, such as 99.99 percent availability of characteristics, without compromising on performance. Cassandra's support for replicating across multiple datacenters is the best in the industry.

Cassandra for DevOps provides comprehensive information on the architecture, installation, monitoring, performance tuning, and configuring Apache Cassandra. You will learn how to address performance and scalability problems and about the tools that can be used for day-to-day administration tasks. Managing a huge amount of data requires a lot of optimization in choosing what to monitor and other management tasks. The simplest administrative tasks can take up a lot of your time; understanding how Cassandra works is the key to spending your time and energy on the right things and making sure you monitor the cluster properly.

What this book covers

Chapter 1, *Basic Concepts and Architecture*, starts by explaining the history and challenges of Big Data and the reason for Cassandra's existence. This chapter explains a bit about the CAP theorem and BigTable concepts.

Chapter 2, *Installing Cassandra*, talks about the hardware choices based on your traffic patterns and YAML configurations. It also talks about EC2 hardware choices as well as and keyspace and column family configurations, including the various options available.

Chapter 3, *Inserting and Manipulating Data*, explains about data modeling. It also talks about creating, altering, and querying data and tracing the query in a column family.

Chapter 4, Administration and Large Deployments, talks about basic administrative tasks, the tools to make data consistent (HH and Repairs), events that can hurt performance, and the ways to reduce the impact.

Chapter 5, Performance Tuning, talks a little bit about things to look out for and tuning for better performance, because the hardest part of running a Java app is getting the right GC settings and dealing with GC pauses. It also talks about request tracing to look for bottlenecks as well as tools to monitor Cassandra.

Chapter 6, Analytics, talks about Hadoop integration and segregating traffic for different work loads.

Chapter 7, Security and Troubleshooting, talks about the various security methods that can be implemented to secure the data in flight and at rest.

Bonus Chapter, Evolving Apps and Use Cases, talks about the various evolving apps and use cases and their tradeoffs. This chapter tries to provide you with information so that you can understand the pitfalls. This chapter can be found online at `https://www.packtpub.com/sites/default/files/downloads/bonus_chapter.pdf`.

What you need for this book

- Sun JVM 1.6 or above
- A Debian or Linux machine
- Basic Internet access
- Enough disk and memory to run Cassandra

Who this book is for

This book is for Cassandra administrators and DevOps responsible for automating and monitoring Cassandra.

Performance engineers and developers who would like to understand the administration of Cassandra will also benefit from this book.

Conventions

In this book, you will find a number of styles of text that distinguish between different kinds of information. Here are some examples of these styles, and an explanation of their meaning.

Code words in text are shown as follows: "The `Cassandra.yaml` file is well documented and is self-explanatory."

A block of code is set as follows:

```
- class_name: org.apache.cassandra.locator.SimpleSeedProvider
parameters:
- seeds: "<IP ADD OF SEED>,<IP ADD OF SEED>"
```

Any command-line input or output is written as follows:

```
Device: rrqm/s wrqm/s    r/s     w/s rMB/s wMB/s avgrq-sz avgqu-sz await
svctm %util
dm-0      0.00    0.00  0.00    3.17  0.00  0.02    10.11     0.01  2.11
2.11  0.67
```

New terms and **important words** are shown in bold. Words that you see on the screen, in menus or dialog boxes for example, appear in the text like this: "If the threads try to create any object larger than the space available for the generations allocated, it will be created directly in **Old Generation**."

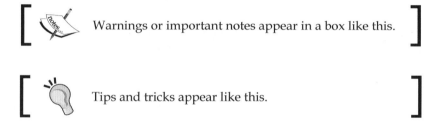

Warnings or important notes appear in a box like this.

Tips and tricks appear like this.

Reader feedback

Feedback from our readers is always welcome. Let us know what you think about this book—what you liked or may have disliked. Reader feedback is important for us to develop titles that you really get the most out of.

To send us general feedback, simply send an e-mail to `feedback@packtpub.com`, and mention the book title via the subject of your message.

If there is a topic that you have expertise in and you are interested in either writing or contributing to a book, see our author guide on `www.packtpub.com/authors`.

Customer support

Now that you are the proud owner of a Packt book, we have a number of things to help you to get the most from your purchase.

Errata

Although we have taken every care to ensure the accuracy of our content, mistakes do happen. If you find a mistake in one of our books—maybe a mistake in the text or the code—we would be grateful if you would report this to us. By doing so, you can save other readers from frustration and help us improve subsequent versions of this book. If you find any errata, please report them by visiting http://www.packtpub. com/submit-errata, selecting your book, clicking on the **errata submission form** link, and entering the details of your errata. Once your errata are verified, your submission will be accepted and the errata will be uploaded on our website, or added to any list of existing errata, under the Errata section of that title. Any existing errata can be viewed by selecting your title from http://www.packtpub.com/support.

Piracy

Piracy of copyright material on the Internet is an ongoing problem across all media. At Packt, we take the protection of our copyright and licenses very seriously. If you come across any illegal copies of our works, in any form, on the Internet, please provide us with the location address or website name immediately so that we can pursue a remedy.

Please contact us at copyright@packtpub.com with a link to the suspected pirated material.

We appreciate your help in protecting our authors, and our ability to bring you valuable content.

Questions

You can contact us at questions@packtpub.com if you are having a problem with any aspect of the book, and we will do our best to address it.

1
Basic Concepts and Architecture

The Apache Cassandra database is a linearly scalable and highly available distributed data store which doesn't compromise on performance and runs on commodity hardware. Cassandra's support for replicating across multiple datacenters / multiple discrete environments is the best in the industry. Cassandra provides high throughput with low latency without any single point of failure on commodity hardware.

Cassandra was inspired by the two papers published by Google (*BigTable*) in 2006 and Amazon (*Dynamo*) in 2007, after which Cassandra added more features. Cassandra is different from most of the NoSQL solutions in a lot of ways: the core assumption of most of the distributed NoSQL solutions is that **Mean Time Between Failures** (**MTBF**) of the whole system becomes negligible when the failures of individual nodes are independent, thus resulting in a highly reliable system.

CAP theorem

If you want to understand Cassandra, you first need to understand the CAP theorem. The CAP theorem (published by Eric Brewer at the University of California, Berkeley) basically states that it is impossible for a distributed system to provide you with all of the following three guarantees:

- **Consistency**: Updates to the state of the system are seen by all the clients simultaneously

- **Availability**: Guarantee of the system to be available for every valid request

- **Partition tolerance**: The system continues to operate despite arbitrary message loss or network partition

Cassandra provides users with stronger availability and partition tolerance with tunable consistency tradeoff; the client, while writing to and/or reading from Cassandra, can pass a consistency level that drives the consistency requirements for the requested operations.

BigTable / Log-structured data model

In a BigTable data model, the primary key and column names are mapped with their respective bytes of value to form a multidimensional map. Each table has multiple dimensions. **Timestamp** is one such dimension that allows the table to version the data and is also used for internal garbage collection (of deleted data). The next figure shows the data structure in a visual context; the row key serves as the identifier of the column that follows it, and the column name and value are stored in contiguous blocks:

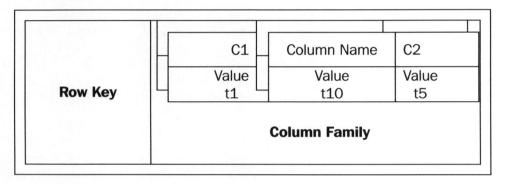

It is important to note that every row has the column names stored along with the values, allowing the schema to be dynamic.

Column families

Columns are grouped into sets called **column families**, which can be addressed through a row key (primary key). All the data stored in a column family is of the same type. A column family must be created before any data can be stored; any column key within the family can be used. It is our intent that the number of distinct column families in a keyspace should be small, and that the families should rarely change during an operation. In contrast, a column family may have an unbounded number of columns. Both disk and memory accounting are performed at the column family level.

Keyspace

A keyspace is a group of column families; replication strategies and ACLs are performed at the keyspace level. If you are familiar with traditional RDBMS, you can consider the keyspace as an alternative name for the schema and the column family as an alternative name for tables.

Sorted String Table (SSTable)

An **SSTable** provides a persistent file format for Cassandra; it is an ordered immutable storage structure from rows of columns (name/value pairs). Operations are provided to look up the value associated with a specific key and to iterate over all the column names and value pairs within a specified key range. Internally, each SSTable contains a sequence of row keys and a set of column key/value pairs. There is an index and the start location of the row key in the index file, which is stored separately. The index summary is loaded into the memory when the SSTable is opened in order to optimize the amount of memory needed for the index. A lookup for actual rows can be performed with a single disk seek and by scanning sequentially for the data.

Memtable

A **memtable** is a memory location where data is written to during update or delete operations. A memtable is a temporary location and will be flushed to the disk once it is full to form an SSTable. Basically, an update or a write operation to Cassandra is a sequential write to the commit log in the disk and a memory update; hence, writes are as fast as writing to memory. Once the memtables are full, they are flushed to the disk, forming new SSTables:

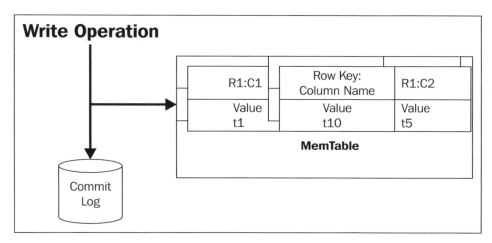

Reads in Cassandra will merge the data from different SSTables and the data in memtables. Reads should always be requested with a row key (primary key) with the exception of a key range scan.

When multiple updates are applied to the same column, Cassandra uses client-provided timestamps to resolve conflicts. Delete operations to a column work a little differently; because SSTables are immutable, Cassandra writes the tombstone to avoid random writes. A tombstone is a special value written to Cassandra instead of removing the data immediately. The **tombstone** can then be sent to nodes that did not get the initial remove request, and can be removed during GC.

Compaction

To bound the number of SSTable files that must be consulted on reads and to reclaim the space taken by unused data, Cassandra performs **compactions**. In a nutshell, compaction compacts n (the configurable number of SSTables) into one big SSTable. They start out being the same size as the memtables. Therefore, the sizes of the SSTables are exponentially bigger when they grow older. Cassandra also supports leveled compaction, which we will discuss in later chapters:

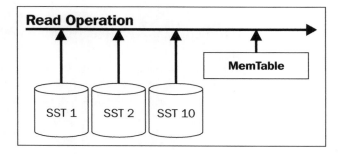

Partitioning and replication Dynamo style

As mentioned previously, the partitioner and replication scheme is motivated by the Dynamo paper; let's talk about it in detail.

Gossip protocol

Cassandra is a peer-to-peer system with no single point of failure; the cluster topology information is communicated via the Gossip protocol. The **Gossip protocol** is similar to real-world gossip, where a node (say B) tells a few of its peers in the cluster what it knows about the state of a node (say A). Those nodes tell a few other nodes about A, and over a period of time, all the nodes know about A.

Distributed hash table

The key feature of Cassandra is the ability to scale incrementally. This includes the ability to dynamically partition the data over a set of nodes in the cluster. Cassandra partitions data across the cluster using consistent hashing and randomly distributes the rows over the network using the hash of the row key. When a node joins the ring, it is assigned a token that advocates where the node has to be placed in the ring:

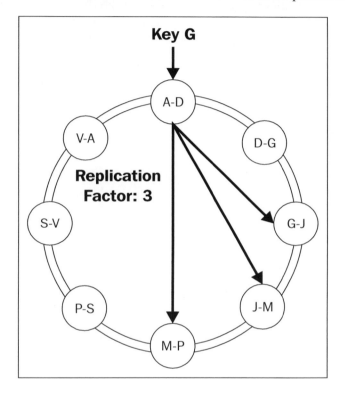

Now consider a case where the replication factor is 3; clients randomly write or read from a coordinator (every node in the system can act as a coordinator and a data node) in the cluster. The node calculates a hash of the row key and provides the coordinator enough information to write to the right node in the ring. The coordinator also looks at the replication factor and writes to the neighboring nodes in the ring order. More on vnodes and multi-DC clusters will be discussed in later chapters.

Eventual consistency

Given a sufficient period of time over which no changes are sent, all updates can be expected to propagate through the system and the replicas created will be consistent. Cassandra supports both the eventual consistency model and strong consistency model, which can be controlled from the client while performing an operation.

Cassandra supports various consistency levels while writing or reading data. The consistency level drives the number data replicas the coordinator has to contact to get the data before acknowledging the clients. If $W + R > Replication\ Factor$, where W is the number of nodes to block on write and R the number to block on reads, the clients will see a strong consistency behavior:

- ONE: R/W at least one node
- TWO: R/W at least two nodes
- QUORUM: R/W from at least *floor (N/2) + 1*, where N is the replication factor

When nodes are down for maintenance, Cassandra will store hints for updates performed on that node, which can be delivered back when the node is available in the future. To make data consistent, Cassandra relies on hinted handoffs, read repairs, and anti-entropy repairs. We will talk about them in detail in later chapters.

Summary

In this chapter, we have discussed basic concepts and basic building blocks, including the motivation in building a new datastore solution. In the next few chapters, we will learn more about them in detail.

2
Installing Cassandra

I would like to start this chapter by showing you some numbers published by Jeff Dean (a Google fellow, `http://www.cs.cornell.edu/projects/ladis2009/talks/dean-keynote-ladis2009.pdf`):

Operation	Time taken	
Send 1K bytes over 1 Gbps network	10,000 ns	0.01 ms
Read 4K randomly from SSD*	150,000 ns	0.15 ms
Read 1 MB sequentially from memory	250,000 ns	0.25 ms
Round trip within the same datacenter	500,000 ns	0.5 ms
Read 1 MB sequentially from SSD*	1,000,000 ns	1 ms
Disk seek	10,000,000 ns	10 ms
Send packet CA->Netherlands->CA	150,000,000 ns	150 ms

The preceding table tells us the average cost of a system call performed to complete an operation. Typically, a read/write request in Cassandra involves multiple of the above operations.

Memory, CPU, and network requirements

To understand the memory requirements for Cassandra, it's important to know that Cassandra is a Java-based service that uses a JVM heap to create temporary objects. Cassandra also uses the heap for its in-memory data structures. Cassandra relies on the OS kernel to manage the page cache of the frequently used file blocks. Most OS kernels have intelligent (multiple) ways to figure out the block of the files that will be accessed by the application and those that can be evicted from its cache.

There are two main functions of any Cassandra node: one is to coordinate the client requests and the other to serve data. The coordinator is a simple proxy, which sends data requests or updates to the nodes that have data and waits for their responses. To achieve quorum, it waits for the $N/2 + 1$ nodes, or it waits for the required nodes as per the consistency levels. Every node in the cluster handles both of these functions; the coordinator contains the most recent information about the cluster via gossip.

Cassandra in-memory data structures

To understand how Cassandra utilizes the available system memory, it is important to understand its in-memory data structures. Cassandra's in-memory data structures are as follows.

Index summary

It is expensive to have the whole index in memory because of its size. Index is a map of row keys and SSTable locations where the actual data resides.

Bloom filter

A bloom filter is the in-memory structure per SSTable, which helps Cassandra avoid a disk seek before the index is scanned for the data bloom filter. It is consulted and checked if the row is present. A bloom filter returns a Boolean advice irrespective of whether the data is in the SSTable or not. It can have a false-positive too. If this happens, we will read the SSTable and return without any row, which is fine since it's an optimization. The bloom filter's false-positive chance can be set in the column family level.

Compression metadata

Compression metadata denotes the start of the compressed block. This metadata is required to be maintained for us to uncompress a block starting from the beginning of the block to read the row.

Cassandra (starting from Version 1.2) uses off-heap memory to store bloom filters, index summary (starting from Version 2.0), and compression metadata. Maintaining the whole index in memory will help speed up the reads; we expect OS to maintain this in its file buffer cache. Since the index is accessed fairly often with reads, there is a better chance for OS to keep the file in memory. In my experience, setting the Java heap from 4 to 8 GB and new generation from 1 to 2 GB gives optimal performance.

The best way to find the right configuration is by testing a few of the custom settings; Visual VM (http://visualvm.java.net/) and jmap (part of JVM, http://docs.oracle.com/javase/6/docs/technotes/tools/share/jmap.html) are your friends.

SSDs versus spinning disks

Cassandra's log structure storage is perfect for spinning as well as SSD drives. When data is read or written to a disc drive, the RW head of the disc needs to move to the position of the disk. The seek time can differ for a given disc due to the varying distance from the start point to where the RW head has been instructed to go. Once the disc seek is complete, the following sequential reads/writes will be faster.

A common problem with SSDs is that frequent updates to a location in the disk can cause a shortening of the time for which it can operate reliably; this is called write amplification. Cassandra inherently uses log-structured storage for its persistence, which is perfect for SSDs. SSDs also eliminate the higher seek times seen in traditional drives.

There are a lot of use cases that are good for spinning drives:

- Reads to writes ratio is low
- Entire data set can fit in the page cache

SSDs are good for most use cases if the project can offer it; consider the following cases:

- Lots of updates that in turn translates to lots of I/O required for compaction
- Strict SLA requirements on reads
- Column family needs a lot of I/O for compactions

The filesystem cache minimizes the disk seeks by using the system memory to access the frequently used data sets.

Key cache

Cassandra uses key caches to avoid disk seeks in addition to all other data structures. When a row is looked-up, Cassandra queries multiple SSTables to resolve the row. Caching the key along with an SSTable offset pointer will allow Cassandra to directly go to the location in the file. Since Version 1.2 column indexes are moved into SStable indexes, it is also cached in the key cache. Hence, the reads are exactly one disk seek when cached.

Row cache

Row caches in Cassandra are not query caches; we try to cache the whole row in memory when a query is executed on a row.

Downloading/choosing binaries to install

Install Oracle JRE from `oracle.com`, where you can find the latest JRE to be installed. Download the binaries from `http://cassandra.apache.org/download/`. You have the choice of installing a Ubuntu or tar package from the site. DataStax has an installation package, which is distributed at `http://www.datastax.com/download`.

Follow the instructions and get it installed; once installed, you will find the following files in the respective locations:

- **Configuration directory**:`/etc/cassandra/conf` or `{CASSANDRA_HOME}/conf`
- **Cassandra packaged installs**: `/etc/cassandra/conf`
- **Cassandra binary installs**:`<install_location>/conf`
- **DataStax Enterprise packaged installs**:`/etc/dse/cassandra`

Configuring cassandra-env.sh

It is advisable to configure the `cassandra-env.sh` to modify:

- `#MAX_HEAP_SIZE="4G"`
- `#HEAP_NEWSIZE="800M"`

It takes a lot of tests to get the GC configurations right, but as a rule of thumb, the default configuration works well for most use cases. We have seen that the following configuration is much more conservative if you have enough memory in the system:

- `#MAX_HEAP_SIZE="8G"`
- `#HEAP_NEWSIZE="1G"`

Enable GC logging and look for concurrent mode failures. Visual GC will provide us with visualizations to understand the GC cycles. Here, the idea is to make sure that most of the garbage is collected in the initial stages, where it is less expensive to be collected. Tuning GC is complicated and a long topic by itself; it is out of the scope of this book too. But testing constantly with various settings at your peek traffic will help mitigate any surprises.

Configuring Cassandra.yaml

The `Cassandra.yaml` file is well documented and is self-explanatory sometimes. But we will talk about a few items in the YAML, which will help users in configuring Cassandra.

cluster_name

cluster_name is the name of the cluster. Cassandra, checks the cluster name before it starts any gossip communication with its peer's nodes.

seed_provider

The seed node has no purpose other than bootstrapping the gossip process for new nodes joining the cluster. Seed nodes are not used beyond the bootstrapping of nodes. Once bootstrapped, cluster information is stored to disc and will be called. While bootstrapping a node to a live cluster, make sure that the seed node list doesn't contain itself. The following shows a sample YAML configuration.

```
- class_name: org.apache.cassandra.locator.SimpleSeedProvider
parameters:
- seeds: "<IP ADD OF SEED>,<IP ADD OF SEED>"
```

Partitioner

A partitioner distributes rows across nodes in the cluster. Any IPartitioner may be used as long as it is on the classpath. Cassandra supports the following partitioners:

- **Murmur3Partitioner random partitioner**: The row key is hashed to find the position in the cluster.

- **ByteOrderedPartitioner**: This orders rows lexically by key bytes (highly discouraged since this partioner can cause hot spots, where some nodes in the cluster will have a lot more data than the others).

- **RandomPartitioner**: It uses the MD5 algorithm to hash the keys; Murmur3Partitioner is faster than MD5. RandomPartitioner is used for backward compatibility and it is deprecated.

auto_bootstrap

Enabling this configuration makes the new Cassandra nodes automatically stream the right data to themselves provided that the cluster is already up and running. Cassandra will ignore this setting when the node is configured as a seed.

You can also re-bootstrap an existing node by deleting all the data directories and starting the node with `-Dcassandra.replace_token=<Token's>` (you can start Cassandra with the `-D` option with the startup script). Essentially, the node that is part of the writing goes into the bootstrap mode and streams the required data into itself, and once it is online, the hints stored for that node are replayed. This setting is very helpful in environments such as EC2, where node recovery is difficult after the node dies (since the data is stored in ephemeral drives). Again, since the advent of vnodes, it is easier to decommission the old node and re-bootstrap the newer node.

broadcast_address

If your Cassandra cluster is deployed in environments where you have two IPs—one for communication within DCs and the other for communication outside DCs—you might want to configure this. The best example for this is AWS, where every instance gets a private and public IP. This configuration needs to point to the public IP for inter-DC configuration.

commitlog_directory

Commit logs usually provide sequential writes to the partition. It is recommended to use a separate disk if available, but it's not always required.

data_file_directories

Make sure to change the name of the data directories. You can specify a list of directories to be used. These directories can be in different physical drives. You can also configure the column family to be in a specific directory. The data directories are configured in the YAML that is to be used. Please note that Cassandra will load balance the data files between different directories, but if you want more control over the way blocks are distributed between the drives, you might want to use hardware RA with some overhead.

There will be cases where it will make more sense to use this configuration when you have a server with both spinning drives and SSD; then, we might be able to configure it.

disk_failure_policy

The default value of this setting is to stop the gossip and thrift when an I/O error or disk failure is detected in order to avoid exposure of the corrupt or bad data to other nodes. You can change it to `best_effort` or set it to `ignore` depending on how you want to deal with errors. It is better to detect errors and stop the node than discovering issues later.

initial_token

`initial_token` assigns the node token position in the ring and a range of data to the node when it first starts up. After vnodes (Version 1.2 and above), this configuration has not been used. Vnodes provide automatic token management. You need to enable the `num_tokens` parameter to control the number of partitions a node will have. Increasing this number results in lots of smaller partitions with a caveat that the range queries on keys will be more expensive.

While using vnodes, there will be more partitions than hosts, and these partitions are distributed across the hosts to create a many-to-one mapping, providing a faster bootstrap and automatic load balancing. Traditional in a distributed hash table style, Cassandra installation of the nodes and tokens has a one-to-one relationship. Hence, while bootstrapping new nodes into the existing cluster, we need to rebalance the cluster, which is an expensive operation. Vnode partitions are randomly distributed across all the nodes.

Losing a node / decommissioning / adding a node to the cluster will evenly redistribute the load to (essentially) all the nodes in the cluster. This drastically reduces the time taken for recovery from failures.

Vnodes work great for almost all cases; but, if you want to disable them for any other reasons, you might want to disable `num_tokens` and fall back to manual token management. There are a lot of tools available online to calculate the tokens and form equally-spaced rings.

listen_address/rpc_address

The IP address of the node is used for communication. If it is left empty, Cassandra tries to find the local hostname/IP to listen on.

Ports

The following are the ports to be configured and their functionalities:

- `rpc_port`: This is the port for the Thrift RPC
- `native_transport_port`: This port is for native transport protocol
- `storage_port`: This is used for internal communication

endpoint_snitch

This sets the snitch that Cassandra uses to place data for safer operation and to avoid data losses during rack or DC failures. The default is a simple snitch that considers the nodes next to each other in different racks and has no notion of DCs. The YAML file has been updated and has more detailed information on what to do. You might need to choose the right one.

commitlog_sync

The writes to Cassandra are being written to the commit log and then to memory, following which the writes to a particular node are considered complete. When writes are performed to a disk partition, the OS caches/batches before it actually performs a disk I/O to write to disk. The application can call fsync to force flush to disk, but this is an expensive process; hence, Cassandra provides two methods of operation:

- **periodic**: The commitlog_sync_period_in_ms method controls when a commit log will be synced to the disk; by default, the sync happens every 10 seconds
- **batch**: This is used with commitlog_sync_batch_window_in_ms to control how long Cassandra waits for other writes before actually doing fsync; writes are not acknowledged until they are fsynced to the disk

commitlog_segment_size_in_mb

This sets the size of the individual commit log segment files. Increase it if you want to reduce the amount of commit log segments. A downside of increasing the number of commit log segments is that it can increase the cost during startup. On the other hand, if you want an incremental backup of every individual transaction on the system, you might want to keep it big enough to allow it to copy to an external source. The commitlog_archiving.properties method allows you to copy the incremental commit log into a separate partition/filesystem for later use. The advantage of commit log backup over incremental Cassandra backup is that it provides all the operations performed in the order of execution on the cluster. The properties file is self-explanatory for most of the properties it supports.

commitlog_total_space_in_mb

This takes care of the total space used for commit log segment files. If the used space goes above this value, Cassandra rotates the next segment. When the space is full and there are no commit log segments available for recycle, memtables are flushed to the disk so that the commit log segments can be reused since they are no longer needed.

It is worth noting that the commit log segments can be recycled once the memtables are flushed to the disk as the data which is in the commit logs. It is already synced to disk. Hence, commit log has redundant information and can safely be thrown away.

Key cache and row cache saved to disk

Cassandra allows you to save the row and key caches to disk, which will enable a restart of the service to bring the caches back to the same state before the service went down.

The `row_cache_size_in_mb` and `key_cache_size_in_mb` settings control the amount of memory used by the row and key caches.

compaction_preheat_key_cache

The key cache has an entry for every SSTable. When compacting, Cassandra creates new SSTables and will invalidate the old ones that were cached. Now, since we have new SSTables, this setting will make Cassandra preheat the rows for the particular row provided the key cache is enabled. This setting is advantageous when a user wants the key cache to be always up-to-date (rather than fault filling). Since compaction can invalidate the keys, it might be worth in some cases to preheat the keys.

row_cache_provider

This configuration is not available in 2.0 and Cassandra uses SerializingCacheProvider. The two types of cache providers are as follows:

- **SerializingCacheProvider**: This serializes the contents of the row and stores it in native memory. Serialized rows take significantly less memory than live rows in JVM. The main idea behind the off-heap cache is that the GC overhead will not be higher as the cache size grows, reducing the stop-the-world GC pauses. Even though the row is stored off-heap, the cache key is stored in Java memory. Hence, there is an overhead in using off-heap cache. You can enable off-heap cache by using `org.apache.cassandra.cache.IRowCacheProvider`. It's worth noting that `row_cache_size_in_mb` only controls the amount of off-heap storage and doesn't limit the heap storage of the cache keys.

- **ConcurrentLinkedHashCacheProvider**: This is purely in-heap cache; you might find this faster in some use cases where the row sizes are small enough to fit in JVM heap memory. `SerializingCacheProvider` performs badly in an update-heavy workload. Since off-heap cache uses native memory to allocate memory and might cause fragmentation of the memory available for future allocation, Cassandra relies on the default allocator to do this (GCC); tests show `jemalloc` has better savings than GCC `malloc`. You might want to try a `memory_allocator` configuration in `Cassandra.yaml`.

Inherently, JVM is not good at managing caches (LRU kind of caches) where the data is frequently freed and new data is added. This causes JVM memory to fragment, causing more GC pressure to compact them. There is no way to hint the JVM about the type of objects.

column_index_size_in_kb

Column index is used to speed up the range query (or queries within a row) on a wide row in order to speed up the lookup/start of a particular column. Increase this setting if your columns K/V are larger. Column index is helpful while doing pagination too. When a query needs to start scanning from the middle of a wide row, Cassandra uses the column index to seek the column faster than scanning the whole row to find the location of the row. As mentioned before, Cassandra's columns inside the rows are always kept sorted, (The order is controlled by the column comparator setting in the column family.) Hence, the index can speed-up reads by directly seeking to the start of the requested data location.

compaction_throughput_mb_per_sec

This is the most important setting, which almost all of Cassandra users modify to get a better I/O performance from Cassandra. This setting throttles compaction disk utilization across all the compactions within the node. The faster the data is inserted, the faster it is compacted to keep the data and SSTable size low enough for the reads to be less expensive. Compaction is mostly sequential with the exception that it has to read from four (default setting) or more SSTs to write a new one. Disabling this setting (by setting it to 0) will take away the valuable I/O bandwidth from the read operations. Some experimentation around the default setting is needed to get the right settings. While using this setting on SSD, you might want to set it to a higher value. With SSD, the limiting factor will be the JVM memory.

in_memory_compaction_limit_in_mb

Cassandra uses a variation of external merge sorts to compact the rows in multiple SSTables to one. This setting controls how much of the memory Cassandra will use in memory. The worst case for this sort is one row that cannot fit in memory. To compact a single wide row, Cassandra then switches to a two-phase compaction, which is expensive. When this happens, a log message is printed; it is suggested to look for these and tune the in-memory size reasonably big enough to accommodate them. But the upper bound of this setting should not surpass 10 to 15 percent of the new generation (GC setting in `Cassandra-env.sh`) as a rule of thumb to avoid GC pressure, hence affecting read or write operations.

concurrent_compactors

This setting will enable Cassandra to do concurrent compaction for multiple column families in the cluster. By default, it is set based on the number of cores in the system. This might be fine in most cases, but there are cases where you need to allow only one compaction in the system to avoid more random reads (reading multiple files and causing more disk seeks). But try tuning `compaction_throughput_mb_per_sec` before trying to optimize the concurrent compactors. All these settings allow Cassandra to be much smarter and safer; most users won't be tuning a lot of these parameters and the defaults work perfectly well.

populate_io_cache_on_flush

When compacting data from old to new SSTs, the filesystem cache is not populated automatically; it takes several cycles before the OS cache is back to normal. This is usually good because the cache is populated organically rather than Cassandra making that decision. But there are use cases where the whole dataset can fit in memory, and we might want to speed up the process of filling the cache. Enabling this setting will force Cassandra to populate the file cache during compaction and flush, hence reducing the spikes in latency (if any due to compaction and filesystem cache). You can confirm this scenario by closely monitoring the I/O wait time of the filesystem.

concurrent_reads

To fetch data from disk, Cassandra employs some reader (worker) threads that read the data from disk (reading BF, SST indexes, and finally reading the data from SSTs), which tells Cassandra how many of them can execute concurrently. This setting enables concurrency for read operations, but at the same time, increasing this number very high will not help read latencies as most of the threads will end up waiting.

As a rule of thumb, make this setting to be 16 multiplied by the number of drives available for reads or writes. For the SSDs per data size that can fit in memory, Cassandra readers will be mostly CPU-bound than disk-bound. For these use cases, set it as a factor of the CPUs available (For example, *8 * NUM_CPU*).

concurrent_writes

The writes in Cassandra are not I/O bound (unless there are disk issues that cause the flush writers to backup). Hence, the number of writer threads should be a factor of the number of CPUs available, and it is recommended to have eight multiplied by the number of CPUs available.

flush_largest_memtables_at

When Cassandra detects that the JVM heap is filling up more than the old objects that JVM GC is able to free, Cassandra tries to free up its reference to memory by flushing the largest memtable to disk. This parameter is an emergency measure to prevent out-of-memory errors. When this happens, Cassandra also logs this incident. When you see this happening, you might want to tune the JVM to avoid it. You might want to look at the `memtable_total_space_in_mb` setting before trying to increase or decrease this setting. This setting doesn't need tuning for most cases.

index_interval

`index_interval` controls the interval at which the index summary is calculated. As mentioned before, an index summary is an in-memory structure for an individual SSTable and allows faster scanning of the SSTable index. Instead of starting the scan from the beginning of the index, Cassandra uses the in-memory structure to quickly jump to the location that is more relevant (skipping others) to start. This setting controls the number of samples that have to be taken to constitute the index summary. Increasing this setting will reduce the memory overhead of the index sample but also increase the amount of sequential I/O required to get to the row. Generally, the best tradeoff between memory usage and performance is a value between 128 and 512. If you are dealing with wide rows and the amount of rows in the cluster is small, consider reducing the interval.

memtable_total_space_in_mb

This is a parameter that controls the total memory used by all the column family's memtables on a given node. This setting needs to be in sync with the used GC heap settings. While using size-tiered compaction, this setting is a key parameter because you might want to flush SSTables slower. Thus, it reduces the number of SSTables to be compacted and provides a little more efficiency with less IO.

memtable_flush_queue_size

This setting controls the number of flush writers that can be queued before backing up the writes. A flush to disk is sequential and you may not end up tuning this setting unless there are some disk issues that cause the flush to slow down.

memtable_flush_writers

This sets the number of memtable flush writer threads. The threads are blocked by disk I/O, and each one holds a memtable in memory while blocked.

stream_throughput_outbound_megabits_per_sec

This throttles the streaming file transfer to the specified MB/s from this node to any other node. Streaming happens when a node is bootstrapped or when it's repaired (after identifying parts of the SSTables which need to be streamed because of inconsistency). Cassandra does sequential I/O to stream data between nodes. Streaming tasks are normally performed during bootstrap, move, rebalance, and repairs. With sequential I/O on the disk, your network becomes a bottleneck, and Cassandra will probably saturate the network cards, and hence throttling the streams will avoid affecting the reads and writes on the cluster.

request_scheduler

This method consists of two components:

- **org.apache.cassandra.scheduler.NoScheduler**: Cassandra doesn't schedule the requests and serves them in the same order as they are received
- **org.apache.cassandra.scheduler.RoundRobinScheduler**: This uses the round-robin algorithm when scheduling the requests with additional options to throttle and add weights on the requests

request_scheduler_options

This method consists of two components:

- `throttle_limit`: This throttles the active requests after the configured number of active requests per client, making sure we are serving all the clients equally. Requests beyond this limit are queued up until running requests have been completed.
- `default_weight`: This denotes the number of requests that will be executed when a client is chosen during the round-robin phase.

rpc_keepalive

This needs to be set when, using a connection, a poll is done in the client side or when the client leaves a connection open for a long time. This is not an absolute necessity if the clients are short-lived, but it doesn't have to be enabled.

rpc_server_type

Cassandra supports two types of RPC servers apart from native services. We highly recommended connection pooling while using a synced RPC server:

- **sync**: This stands for one thread per RPC connection. For a very large number of clients, memory is the limiting factor.

- **hsha**: This stands for a half synchronous, half asynchronous RPC server type. There is a set of threads (based on the number of CPUs) that is used to manage the connections. The HSHA server collects the data from the connections (which is usually CPU-bound) and executes the requests in a RPC thread pool. Basically, the advantage is you don't need to have n threads for n connections like in the RPC server. In the latest versions of Cassandra, this setting might provide a better response time (99 percentile).

Since Version 1.2, Cassandra provides an alternative Netty-based transport implementation instead of thrift. Setting `start_native_transport` can enable native transport.

thrift_framed_transport_size_in_mb

Frame size (maximum field length) for thrift is used to create buffers for incoming and outgoing data.

rpc_max_threads

Both HSHA and sync RPC servers use a thread pool to execute any operation received from the client. This setting sets the maximum number of threads that will be used to execute the operations. Native transport has a similar setting, `native_transport_max_threads`, for its thread pool. Currently, there is an effort to remove these thread pools in the Cassandra 2.1 release and reduce the overhead.

rpc_min_threads

This sets the minimum thread pool size for remote procedure calls. For native transport, use `native_transport_min_threads`.

Timeouts

This is the time in milliseconds for which a coordinating node will wait for the reply from other nodes before the command is timed out. This setting is used intensely by Cassandra internally. Reducing this number to a smaller number might cause trashing of the connections and resources. It needs to be set reasonably higher for the operations to be completed and small enough so the clients can retry in:

- `read_request_timeout_in_ms`: This controls how long the coordinator should wait for read operations to complete

- `range_request_timeout_in_ms`: This controls how long the coordinator should wait for the range scan request to complete

- `write_request_timeout_in_ms`: This controls how long the coordinator should wait for writes to complete

- `truncate_request_timeout_in_ms`: This controls how long the coordinator should wait for truncates to complete

- `request_timeout_in_ms`: This is the default timeout for other miscellaneous operations

- `streaming_socket_timeout_in_ms`: This defaults to 0 (disabled) and denotes the socket timeout for file streaming. When a timeout occurs during streaming, it is retried. When there are network issues, the socket can hang forever. To avoid this situation, set this setting to timeout the connection if the other node doesn't respond on time. Avoid setting this value too low as it can result in a significant amount of data restreaming, which might in turn lead to too many small compactions in the other node.

Dynamic snitch

The community has spent a lot of time tuning these parameters for better performance. Most of the time you don't need to tune them; however, it is nice to know about it:

- `dynamic_snitch_badness_threshold`: This parameter controls the threshold for dynamically routing requests away from a poorly performing node. The percentage tolerance for the latency is set via this parameter; for example, `0.1` means Cassandra will try to avoid nodes whose latencies on operations (overall) compared to other nodes in the cluster is 10 percent worse. Until the threshold is reached, incoming client requests are statically routed to the closest replica chosen by the snitch. This setting is the key to pin all the requests to a particular host. Setting this lower will cause the read operations to be routed indeterminately. Changing this setting to a higher value will cause latency increase when the node to which the request is routed to overloads. During compaction and repair, the severity of those operations is added to the badness threshold, which causes the coordinator of the request to prefer a different node than the one in the list. The idea is to avoid nodes that have a lower chance of completing the request in time.

- `dynamic_snitch_reset_interval_in_ms`: This describes the interval in milliseconds at which the heuristics of the nodes are reset, allowing traffic to the bad nodes to receive traffic again.

- `dynamic_snitch_update_interval_in_ms`: The time interval in milliseconds for calculating the threshold.

- `phi_convict_threshold`: This property controls the sensitivity of the failure detector on an exponential scale. Lower values increase the likelihood of an unresponsive node to be marked as down, but care should be taken while working with environments which have more transient networks and node failures. Any hiccup there can cause all the nodes to be marked down, resulting in a failure in the client side.

- `hinted_handoff_enabled`: This setting enables or disables hinted handoff. As mentioned before, the hinted handoff is one of the mechanisms that keeps the data consistent when the write operation is not performed on a specific node. A hint indicates that the write needs to be replayed to a node that was either down or did not respond to a write request. Hints are spread across the whole cluster for a particular node. When the node comes back online, these hints are replayed by the coordinator to the specific node.

- `max_hint_window_in_ms`, `max_hints_delivery_threads`: This parameter controls the hint window for which the hints are stored; after this, we start to drop the hints. This setting avoids storing a lot of hints for the nodes which won't come back. For a node that is down for a longer period, it might well be cheaper to bootstrap the node than allowing it to replay all the hints.

- `max_hints_delivery_threads`: This will tell Cassandra how many threads have to be used to replay the hints. In a multi-DC Cassandra installation, the hints replayed are naturally throttled by network latencies. There are cases where the throttle causes the nodes to replay hints very slowly; hence, having more threads provides better concurrency.

- `hinted_handoff_throttle_delay_in_ms`: When the coordinating node/ proxy detects that a data node for which it is holding hints has recovered, it begins sending the hints to that data node. This setting specifies the sleep interval in milliseconds after delivering each hint to the destination. This is a time throttle as every node in the cluster is trying to replay and hence a network throttle is irrelevant. For larger deployments, increasing this number will help; for smaller deployments, increasing this number will increase the time for the node to become consistent. It is worth noting that hinted handoff is the best effort to make the data consistent; however, repair the cluster or always bootstrap a downed node if you are worried about consistency. Cassandra also logs messages when it starts to drop hints, and when you see this message in `system.log`, it is time to run the repair when all the nodes are up.

- `inter_dc_tcp_nodelay`: TCP tries to batch the transfer of data from one node to another to avoid small packet transmission. Most of the applications disable this feature because it causes additional latency to every request. But for multi-DC cases where the clusters are geographically distant from each other, this setting helps.

Backup configurations

Out of the box, Cassandra provides incremental, snapshot, and commit log backup configurations. Backup in Cassandra is unique because SSTs are immutable and can safely be backed up. Incremental and snapshot backup operates on SSTables directly by creating a hard link of the files to a separate directory (which is a low overhead operation). External scripts can watch the directory and copy the files out of the node for archival. A snapshot of the nodes can be taken via the `nodetool` command (most users run cron jobs).

incremental_backups

Incremental backup, if enabled, will create hard links to the SSTs once flushed to disk. The idea is that snapshot backup is a full backup of the node, and flush is the new data which is written to Cassandra since it was snapshotted.

Compacted SSTs are not backed up because they are just a copy of the flushed SSTs. To restore data into the cluster, the snapshot has to be restored and incremental SSTs can be replayed on top before starting the nodes.

auto_snapshot

This setting controls whether Cassandra has to take a snapshot before the keyspace or the column family has to be truncated or dropped. This setting is strongly advised and has saved many users from accidentally dropping the CFs via CQL/CLI.

Cassandra on EC2 instance

While writing this book, the following instance types where available in EC2:

Memory	Compute units	Storage	I/O performance	API name
60.50 GB	88	3370 GB (4 x 840 GB)	Very high	cc2.8 x large
23.00 GB	33.5	1690 GB (2 x 840 GB)	Very high	cc1.4 x large
22.00 GB	33.5	1690 GB (2 x 840 GB)	Very high	cg1.4 x large
60.50 GB	35	2048 GB (2 x 1024 GB SSD)	Very high	hi1.4 x large
244.00 GB	88	240 GB SSD	Very high	cr1.8 x large
117.00 GB	35	48 TB (24 x 2 TB)	Very high	hs1.8xlarge
7.00 GB	20	1690 GB (4 x 420 GB)	High	c1. x large
1.70 GB	5	350 GB	Moderate	c1.medium
34.20 GB	13	850 GB	High	m2.2 x large
17.10 GB	6.5	420 GB	Moderate	m2. x large
68.40 GB	26	1690 GB (2 x 840 GB)	High/1000 Mbps	m2.4 x large
15.00 GB	8	1690 GB (4 x 420 GB)	High/1000 Mbps	m1. x large
7.50 GB	4	850 GB (2 x 420 GB)	High/500 Mbps	m1.large
3.75 GB	2	410 GB	Moderate	m1.medium
1.70 GB	1	160 GB	Moderate	m1.small
30.00 GB	26	0 GB (EBS only)	High	m3.2 x large

Memory	Compute units	Storage	I/O performance	API name
15.00 GB	13	0 GB (EBS only)	Moderate	m3. x large
0.60 GB	2	0 GB (EBS only)	Low	t1.micro

(Courtesy: `http://www.ec2instances.info/`)

Multiple tests show that using ephemeral storage drives is better than using EBS volumes for most use cases. Writing to EBS volumes is similar to writing to an NFS and are remote volumes, which can be a bottleneck on network speed and bandwidth. The downside of this approach is that when instances die, the data goes with it; the data in these drives is not recoverable. Users need to make sure the replication factor on the keyspaces is greater than one to avoid data loss.

Looking at the above EC2 instance configuration, users can choose the right instance that fits their needs. If the use case is write heavy, choose an instance with higher disk space; else, go for a higher memory instance for better page cache efficiency.

Snitch

Cassandra provides two snitches to assist in the AWS/EC2 setup:

- **Ec2Snitch**: EC2s are deployed in a single region. The Ec2snitch loads region and availability zone information from the EC2 API. EC2region is treated as the datacenter and the availability zone as the rack. Hence, it is important to start the cluster in multiple availability zones (a minimum of two availability zones is recommended). This setup also increases your cluster's availability when the availability zone fails. In the past, we have seen any one availability zone going down but rarely two.

- **Ec2MultiRegionSnitch**: This uses a public IP as the broadcast address to allow cross-region connectivity. In AWS, the instance cannot listen in a public IP, whereas the traffic to the public IP is automatically redirected to the private IP. The private IP can be addressed only within the region, and it can be addressed across the region using the public IP. Ec2MultiRegionSnitch automatically handles this case and private/public IP information is queried from the EC2 APIs. By default, the security group will not open the ports between the EC2 regions. The EC2security groups are within a region and cannot span across regions. To get around the problem, explicit ACLs (security group settings) need to be set up between the regions to open the storage port / SSL storage port to enable Cassandra nodes to talk. Having said that, the traffic between the regions is not encrypted; by default, set `internode_encryption` to DC in order to enable encryption for the traffic between the DCs.

Create a keyspace

To create a keyspace, we should log in to the CQL shell. It can be found inside the `bin` directory. Help create keyspace on the CQL shell provides self-explanatory help on creating the keyspace:

```
CREATE KEYSPACE<ksname> WITH replication = {'class':'<strategy>'
[,'<option>':<val>]};
```

The CREATE KEYSPACE statement creates a new top-level namespace (also known as a keyspace). Valid names are any strings constructed of alphanumeric characters and underscores. Names that do not work as valid identifiers or integers should be quoted as string literals. Properties such as `replication_factor` and datacenters to replicate are specified during creation as the key-value pairs in the replication map. `'class'` (required) is the name of the replication strategy class which should be used for the new keyspace. Some often-used classes are SimpleStrategy and NetworkTopologyStrategy. Most strategies require additional arguments that can be supplied as key-value pairs in the replication map.

To create a keyspace with NetworkTopologyStrategy and the strategy option of DC1 with a value of 1 and DC2 with a value of 2, you would use the following statement:

```
CREATE KEYSPACE<ksname> WITH replication =
{'class':'NetworkTopologyStrategy', 'DC1':1, 'DC2':2};
```

To create a keyspace with the SimpleStrategy and replication_factor options with a value of 3, you would use the following statement:

```
CREATE KEYSPACE<ksname>WITH replication =
{'class':'SimpleStrategy', 'replication_factor':3};
```

The most important setting concerning creating the keyspace as an administrator is the replication factor. A replication factor is set at the keyspace level, and during the creation of the keyspace, the user needs to decide the number of replicas required. Cassandra also allows users to specify the replication topology at the keyspace level; simple topology ignores the network placement of the nodes and places the replicas next to each other in the token order, whereas NetworkToplogyStategy tries to place the replicas in different racks. It is advisable to use NTS in all the cases so that it will be easier to expand the cluster to a multiregion setup as and when needed.

While configuring a multiregion cluster, the user needs to choose the number of replicas that need to be present in each region. A multiregion replication configuration will look like the following:

```
replication = {'class':'NetworkTopologyStrategy', 'DC1':3,
'DC2':3};
```

The number of replicas depends largely on the use case. If the use case requires the server to have strong consistency within a DC, the application has to do quorum. Hence, the number of replicas needs to be in odd numbers. A replication factor of one will cause the reads to go across the DC when a node is down, in turn causing an increase in latency. Care should be taken while choosing the replication factor.

Creating a column family

To create a column family, you need to log in to cqlsh; the syntax for cqlsh is self-explanatory:

```
cqlsh> help CREATE_TABLE;
      CREATE TABLE <cfname>(<colname><type> PRIMARY KEY [,

<colname><type>[, ...]] )
                [WITH <optionname> = <val> [AND <optionname> =
<val> [...]]];
```

The CREATE TABLE statement creates a new CQL table under the current keyspace. Valid table names are strings of alphanumeric characters and underscores which begin with a letter.

Each table requires a primary key which will correspond to the underlying column family key and key validator. It's important to note that the key type you use must be compatible with the partitioner. For example, OrderPreservingPartitioner and CollatingOrderPreservingPartitioner both require the UTF-8 keys.

In the CQL3 mode, a table can have multiple columns composed of the primary key (execute HELP COMPOSITE_PRIMARY_KEYS in the CQL console). For more information, see one of the following:

```
cqlsh> help CREATE_TABLE_OPTIONS;
CREATE TABLE: Specifying columnfamily options
CREATE TABLE blah (...)
  WITH optionname = val AND otheroption = val2;
```

Any optional keyword arguments can be supplied to control the configuration of a new CQL table. The CQL reference document has a complete list of options and possible values:

```
cqlsh> help CREATE_TABLE_TYPES;
CREATE TABLE: Specifying column types
CREATE ... (KEY <type> PRIMARY KEY,
othercol<type>) ...
```

It is possible to assign a type to the columns during table creation like in traditional databases. Columns configured with a type are validated accordingly when a write occurs, and intelligent CQL drivers and interfaces will be able to decode the column values correctly when receiving writes in the coordinator. Column types are specified as a parenthesized, comma-separated list of column terms and type pairs. Cassandra stores all the columns sorted in disk. If the columns become wide, Cassandra automatically creates a bloom filter and column indexes for efficiency in fetching the columns. Having said this, comparator is the most important part of the column family settings, which advices Cassandra to sort columns in a particular order. This setting is an immutable setting; once set, it cannot be changed.

By default, there are multiple comparator types available for column names:

```
cqlsh> HELP TYPES;
```

The preceding command on CQL should list the following data types: `Ascii`, `bigint`, `blob`, `Boolean`, `counter`, `decimal`, `double`, `float`, `inet`, `int`, `list`, `map`, `set`, `text`, `timestamp`, `timeuuid`, `uuid`, `varchar`, and `varint`.

In the preceding list of recognized types, maps and lists are exceptions. CQL tries to handle them using composite column constructs. Internally, the map's key is merged with the column name as a composite column, and the list's index is merged into the column name.

For information on the various recognizable input formats for these types or controlling the formatting of `cqlsh` query output, see one of the following topics:

```
cqlsh>HELP TIMESTAMP_INPUT
cqlsh>HELP BLOB_INPUT...
```

GC grace period

As mentioned earlier, tombstones are the deletion markers that are written to the SSTable when the delete commands are executed. Read and manual repairs try to keep the Cassandra replicas in sync by comparing the same data from different servers. (Inconsistency can happen due to a minor hiccup, network partition, or hardware failures.) If Cassandra detects an inconsistency, data is copied and made consistent.

If the tombstone is not stored long enough and a node misses the delete, the deleted column can be resurrected from the node that has inconsistent data. To avoid this situation, we need to hold the tombstone in Cassandra long enough so that all the nodes receive the deletion markers.

Wait in seconds before the tombstones are garbage collected; the default value is 10 days.

Compaction

Cassandra supports two kinds of compaction out of the box (there are additional compaction strategies available), `SizedTieredCompactionStrategy` and `LeveledCompactionStrategy`.

Cassandra's size-tiered compaction strategy is similarly described in Google's Bigtable paper. When enough similar-sized SSTables are present, Cassandra will merge them to one bigger file; in other words, the compaction is similar to an external merge sort.

Leveled compaction creates SSTables of a fixed size and in small chunks which are grouped into levels. Within each level, SSTables will not overlap. `sstable_size_in_mb` controls the size of the SSTable in the first level. Increasing the default size, in most cases, has a better performance and memory usage. Each level will be ten times as large as the previous level. New SSTables are added to L0 (level 0) and immediately compacted with the SSTables in L1. When L1 fills up, extra SSTables are promoted to L2. Subsequent SSTables generated in L1 will be compacted with the SSTables in L2 with which they overlap.

There are a lot of advantages in leveled compaction. Ninety percent of the reads can be performed with a read to one SSTable; only 10 percent of the size needs to be reserved for compaction (whereas 50 percent needs to be free for tiered compaction). The biggest disadvantage is that leveled compaction requires more I/O throughput.

Minimum and maximum compaction threshold

The minimum number of SSTables needed to trigger/schedule a compaction defaults to four. The maximum value is the maximum number of SSTables that are allowed before forcing compaction, which defaults to 32. Increasing the minimum/maximum value will reduce the frequency of compaction, and reducing the minimum/maximum value will increase the frequency of compaction.

In some cases, tuning these settings helps, especially when the server is starving for I/O or most of the column families have short TTLs, and you would like Cassandra to drop them without merging them.

Secondary indexes

Cassandra supports secondary indexes. Without secondary indexes, users might have to query all the data (key range queries) or use MapReduce hive queries. A secondary index is nothing but a reverse index of `RowKey:ColumnName:Value`. In other words, indexes on column values are called secondary indexes.

Secondary indexes allow querying by value and can be built in the background automatically without blocking reads or writes. Cassandra implements secondary indexes as a hidden table, separate from the table that contains the values being indexed. While creating the column family or later (using the `alter table` syntax), the user can specify the column names to be indexed by Cassandra using the following queries:

```
CREATE INDEX name ON address (name_idx);
SELECT * FROM address WHERE name_idx = 'VJ';
```

One thing to note is that the secondary index in Cassandra, as of Version 1.2, does read before write (read the index and update the index); hence, adding a secondary index will slow the writes and take more I/O.

Composite primary key type

Also known as a composite column, this primary key type allows users to have multiple types of column names. Prior to Version 1.0, super columns provided flexibility for users to organize columns so they can be viewed and queried as a column of columns, that is, an additional dimension to represent a group of columns. Super columns have their own drawbacks, and reading them causes the whole super column to be loaded into memory. Composite columns are a better way to represent a hierarchy of columns.

```
UUID: {  Address: {  "City": "San Jose",
  "State": "California",
  "Street": "ABC Road" }
Twitter: {  "handle": "@vijay2win" } }
```

The composite primary key in CQL uses composite columns where the first part of the composite primary key is used as a row key. The second part of the primary key is represented as the first part of the column name. The following are the available options when creating or modifying a column family:

- **key_validation_class**: This details the primary key type that will be written to and read from. Cassandra allows users to set this purely for data validation. This setting is required when using secondary indexes.

- **default_validator**: As the name suggests, this is the default validator for the values inserted into the column family.

- **Column metadata**: Metadata about the column, usually the validation class for the values, is useful when the column has multiple types of values for corresponding column names and they have to be validated individually. This setting is more useful in cases of secondary indexes and CQL queries.

- **Speculative reads**: The documentation in `cassandra-cli` is self-explanatory; when the requests are sent to the replica and the read-repair chance is lower than 1.0, fewer nodes are consulted to complete the read request. When the nodes become slower, the overall latency of the read requests increases. To avoid this situation, Cassandra also supports speculative reads that have the configuration to speculate additional read requests to retry the read request on a different node.

```
ALTER TABLE users WITH speculative_retry = '15ms'
```

  Following are the options available to you:

    - `xms`: As shown previously, this allows you to specify x retry time based on the SLAs

    - `xpercentail`: This will advice Cassandra to retry if a read request doesn't complete within the x percentile of observed latencies from the past.

    - `Always`: This will advice Cassandra to retry always.

- **Caching**: This allows users to choose the kind of caching that has to be performed on the data in a column family.

> The limits to the cache memory are performed in `Cassandra.yaml` to specify the maximum memory size to be used for each type of cache; this setting is a directive to Cassandra to allocate memory. If caching is disabled in YAML, this setting is ignored.

- **Compact storage**: Cassandra enables compression by default. For performance reasons, if you have to disable compression, store `null` on `sstable_compressor` when using compression.

Options

The following are the options available while configuring the column family:

- **sstable_compression**: This is the compression algorithm to be used.
- **chunk_length_kb**: This is the block size (defaults to 64 Kb). A configured block of data is compressed and written to disk. A block index is kept in memory for every block of data written for SSTable, and an in-memory block index allows a faster scan into the block.
- **crc_check_chance**: For every block of data written to disk, a CRC is calculated and written along with the data; during decompression, the CRC is verified to identify corruption. CRC calculation and comparison is CPU-intensive and this setting provides a tradeoff.

Since the column names are repeated in a big table / columnar storage, the bigger the block size, the better the compression ratio. The tradeoff is that reads have more work to do in decompressing large blocks.

read_repair_chance and dclocal_read_repair_chance

These are probabilities (0.0 to 1.0) to perform read repairs against the incoming read requests. If `dclocal_read_repair_chance` is lower than `read_repair_chance`, the global settings will precede. `dclocal` read repair makes more sense on multi-DC clusters. This setting will reduce the network bandwidth between the DC and load on the nodes in remote DCs (the DC that received the read in this context is considered local), keeping the node in the local DC more consistent.

For instance, *read_repair_chance = 0.1* and *dclocal_read_repair_chance = 0.5*.

For 10 read queries, one will do read repair on all replicas; four will only do read repair on the replica of the local DC, and five will not do any read repair.

Summary

In this chapter we discussed how to install and configure Cassandra at a high level. In the following chapters, we will cover more about the data structures that Cassandra is based on and the effectiveness of multiple configurations. In the next few chapters you will be exposed to more in-depth details on how Cassandra fetches data from disk and how it optimizes itself, helping us understand the implications on data modeling, maintaining, and troubleshooting.

3

Inserting Data and Manipulating Data

Datastax has an updated reference documentation at `http://www.datastax.com/docs/1.1/references/cql/index` which we would like to suggest referring to in case you need more than what we cover here. We will be covering things which will be useful as an administrator.

Querying data

Cassandra supports multiple interfaces for clients to query data; native transport supports Cassandra Query Language queries (with clients written in multiple languages). With the help of Thrift, Cassandra supports multiple languages for which native transport support doesn't exist. **Cassandra Query Language (CQL)** is similar to SQL and is used for querying Cassandra. In other words, CQL is a **domain-specific language (DSL)** for querying data from Cassandra. You can also choose to use the `cqlsh` command to connect to Cassandra to run ad hoc CQL queries.

USE

A USE statement sets the keyspace of the current client session; all subsequent operations in the connection are within the context of the keyspace. You can reset the keyspace by executing the USE command again:

```
USE <keyspace_name>;
```

CREATE

CREATE TABLE creates a new column family under the current keyspace.
While creating a table, we need to define the primary/row key's data type and the comparator for the column family. In CQL3, the column family type is assumed to be string and the value types can be defined. You can log in to CQL to define a different data type to the column names if you wish.

In case of a composite primary key (as shown in the following code snippet), Cassandra uses the first part of the primary key as the actual partition key; the second part of the primary key is used as the first part of the composite column name, following which we will have the actual column family name defined:

```
CREATE TABLE <column family name>
(
<column_name>        <type>,
[<column_name2>      <type>, ...]
PRIMARY KEY
(
<column_name>        <type>
[, <column_name2>    <type>,...]
)
[WITH <option name> = <value>
[AND <option name> = <value> [...]]]
);
```

Cassandra also supports **secondary indexes**, but the index queries span across the cluster; this will be cheaper on a smaller installation and expensive on large installations. Cassandra secondary indexes are a good fit for low cardinality values (more unique values):

```
CREATE INDEX table ON users (<Column name>);
```

ALTER

ALTER TABLE is used to alter the table schematics after the table or keyspace is created and used. The schematics for the command are the same as CREATE, except the keyword needs to be replaced with ALTER.

Comparator or the primary key cannot be changed. In case a user needs to change the comparators, they need to recreate the column family and repopulate the data. Dropping schematics of the column doesn't actually drop it from the filesystem (unlike RDBMS), it just becomes invisible.

```
ALTER TABLE [<keyspace_name>].<column_family>
(
ALTER <column_name> TYPE <data_type>
| ADD <column_name><data_type>
| DROP <column_name>
| WITH <optionname> = <val> [AND <optionname> = <val> [...]]
);
```

DESCRIBE

The DESCRIBE command is used to describe an object's valid options: DESCRIBE CLUSTER, DESCRIBE TABLE, DESCRIBE KEYSPACE, and DESCRIBE SCHEMA. It is highly recommended that you run these commands on your test cluster while reading this section.

The DESCRIBE CLUSTER option allows users to view the cluster name, partitioner, and snitch information.

The DESCRIBE SCHEMA option allows users to view schema information. You can copy-paste to replicate the schema.

SELECT

Selecting data out of Cassandra will follow the following syntax:

```
SELECT <select expression> FROM <column family>
[USING CONSISTENCY <level>]
[WHERE (<clause>)] [LIMIT <n>]
[ORDER BY <composite key 2>] [ASC, DESC]
<clause> syntax is:
<primary key name> { = | < | > | <= | >= } <key_value>
<primary key name> IN (<key_value> [,...]);
```

It is important to understand that an efficient query with a WHERE clause should be specified with a primary key. The first part of the primary key query has to be an exact match and the second part can have <, >, or = clauses (primary key order does matter).

Tracing

For an administrator, the most valuable feature in Cassandra is tracing query performance. If you are familiar with RDBMS, you might remember EXPLAIN <query>, which gives detailed information about the query. Similarly, in Cassandra, you can trace any query. Users can trace by enabling the following command in cqlsh:

```
cqlsh:Keyspace1> tracing on;
Now tracing requests.
cqlsh:Keyspace1> select key from "Standard1" limit 1;
key
--------------------
 0x3035373736343535
Tracing session: 19731600-9a75-11e2-91b2-89ac708f7e3b

activity | timestamp     | source  | source_elapsed
----------------------------------------------------------
execute_cql3_query | 19:37:30,553 | 17.2.1.1 | 0
Parsing statement | 19:37:30,553 | 17.2.1.1 | 44
Preparing statement | 19:37:30,553 | 17.2.1.1 | 246
Determining replicas to query | 19:37:30,553 | 17.2.1.1 | 369
Message received from /17.2.1.1 | 19:37:30,554 | 17.2.1.2 | 52
Sending message to /17.2.1.2 | 19:37:30,554 | 17.2.1.1 | 910
Executing seq scan across 13 sstables for [min(-9223372036854775808),
max(-2635249153387078804)] | 19:37:30,554 | 17.2.1.2 |    358
Scanned 1 rows and matched 1 | 19:37:30,557 | 17.2.1.2 | 3649
Enqueuing response to /17.2.1.1 | 19:37:30,558 | 17.2.1.2 | 3692
Sending message to /17.2.1.1 | 19:37:30,558 | 17.2.1.2 | 3834
Message received from /17.2.1.2 | 19:37:30,762 | 17.2.1.1 | 208985
Processing response from /17.2.1.2 | 19:37:30,762 | 17.2.1.1 | 209111
Request complete | 19:37:30,762 | 17.2.1.1 | 209295
```

Let's explain the preceding output so we can understand how tracing works:

- The first three statements show how long the CQL parsing takes
- The next three lines show how long it took to send the request to the right node to execute the requests
- The seventh line shows how many SSTables were scanned to get the data out and shows their corresponding latencies
- The ninth to eleventh lines show the messages that were sent back to the coordinator

If you see unusually long time consumption in any of the stages, it is worth troubleshooting or tuning that part of Cassandra. As you can see, tracing in Cassandra is self-explanatory, and it exposes a lot of information to help you troubleshoot.

When the JMX metrics about latencies, threads waiting, and so on doesn't help, try tracing. It is important to note that there is a huge performance penalty on enabling tracing on every (or production) request, and should only be used in troubleshooting.

Data modeling

Now that we know a bit more about Cassandra and its architecture, let's dive into data modeling. Cassandra supports both dynamic and static/fixed column names:

- **Fixed columns**: Similar to the RDBMS table structure, Cassandra defines (explicitly created during the table creation), validates, and inserts columns ahead of time. Secondary indexes can also be created on them. There are cases where this model has advantages, such as where the data store is shared between multiple applications and the data needs to be validated before it is inserted.

> The column names still need to be of the same type.
> The Cassandra data storage structure is very different from traditional databases; when a column is not there, instead of storing a null value, Cassandra doesn't store the column at all.

```
CREATE TABLE users (
user_idint PRIMARY KEY,
name varchar,
address varchar,
rank int,
score int
 );
```

- **Dynamic columns**: The power of Cassandra is the BigTable data structure where any row in the database can have millions of unique column names. This structure allows numerous use cases where the column name of the incoming data is not known ahead of time.

```
CREATE TABLE users (
user_id   int PRIMARY KEY,
activity map<timestamp, text>
 );
```

Types of columns

Cassandra supports multiple types of columns to represent your data; they are described as follows:

- **Standard columns**: This can be addressed with the primary key / row key. A standard column also contains a timestamp attached to it, which allows Cassandra to perform conflict resolution. The column family is a set of columns addressed by a primary key, for example, `Select * from table where key="xyz";`.

- **Composite columns**: This is similar to a standard column family with additional dimensions on the data allowing the user to query/address the set of columns. For example:

```
Select * from table where key="xyz" and col1 > "abc" and col1 <
"xyz";
CREATE TABLE library (
user_id      varchar,
checkout     timestamp,
title        varchar,
author       varchar,
desc         varchar,
  PRIMARY KEY (user_id, author)
);
```

  Example query:

```
INSERT INTO crew (user_id, checkout, title_author) VALUES ('john_
doe', '2013-09-12', 'Things Fall Apart', 'Chinua Achebe');
SELECT * FROM library WHERE user_id='john_doe' AND started_at>
'2013-09-11';
```

- **Expiring columns**: Traditional databases need separate external scripts to clean up the data that needs to expire. In Cassandra, we can specify the time a column should live; when added, the column expires or tombstones automatically once the time reaches the limit without any additional action required.

```
UPDATE library USING TTL=6800 …
```

- **Counter columns**: This is a column used to store a number that incrementally counts the occurrences of a particular event. Typical use cases for counters are distributed counts of a particular application metric, number of retweets on a particular tweet, number of followers, and so on. In Cassandra, counters take a separate path in execution to ensure they do not overcount during failures; hence, the user needs to configure counter column families separately from the normal column families. Counters also perform read operations before writing.

```
UPDATE ... SET name1 = name1 + <value> ...;
```

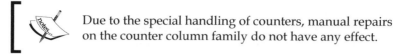

> Due to the special handling of counters, manual repairs on the counter column family do not have any effect.

Common Cassandra data models

Let's talk about some of the most common Cassandra use cases and data models. Cassandra doesn't support complex joins, and hence, as a user, it is important to look at the read queries and then model the data accordingly. In RDBMS, you would try to normalize the data so you don't have duplicates, but in a NoSQL solution like Cassandra, you would denormalize the schema to achieve effective read queries.

Denormalization

In the relational world, the data model is usually designed upfront with the goal of normalizing the data to minimize redundancy. In RDBMS, the idea is to create smaller, non-redundant data structures that are connected to each other by foreign keys and other constraints.

Cassandra does not have constraints or foreign key relationship out of the box, which means that users have to deal with it. Don't panic, we are not pushing the complexity to the users. NoSQL systems were introduced because of the inherent complexity in the RDBMS systems, which slow down performance at scale. If you have strong use cases that can be solved in RDBMS with relationships, you should consider using RDBMS.

Denormalization is a process in which you look back at the user queries and structure the data accordingly. Writes are cheaper in Cassandra, hence writing more for faster query performance is absolutely fine. You might want to think about where denormalization makes sense. Denormalization does take more space. Cassandra queries are faster than traditional RDBMS too, querying more doesn't hurt. Most users tradeoff between reading more versus denormalizing/adding more dimensions to the data.

To put this into perspective, most RDBMS-based applications that have serious data already perform denormalization. Consider a case where the data has 20 relationships and the RDBMS has to query 20 places to construct a result that causes 20 random IO operations and naturally slows down the performance; if you denormalize, you can get around this problem with a fewer disk seeks.

- **Time series database example**: Cassandra is an excellent fit for time series data. The following list, although not exhaustive, contains some examples of TSDB use cases:
    - ° Ratings, recent purchases, and shopping cart
    - ° Session data, event streams, and click streams
    - ° Sensor data, application, and performance metrics
    - ° Velocities or windowed queries for a specific time period

- **Comparator**: One of the biggest decisions that has to be taken is choosing the comparator for the time series data. You can choose to store the timestamp as long as there won't be concurrent updates or overrides to the same column. A classic example is that of application metrics being updated at constant intervals. For most cases, you might want to use TimeUUIDs that are guaranteed not to overwrite each other (most clients do some interesting tricks to guarantee uniqueness within a node). You can also choose string with Joda-Time; by now you might have understood the reason for this. Column comparators can be reversed; for example, TimeUUIDs as column names can be stored in disks in the reverse order, where the latest data is stored at the start of the row (reverse chronological order) and can improve the performance a bit if your query always looks for the latest dataset.

    With time series datasets, you are building an index. You first fetch a set of row keys from a timeline and then multiget (batch get on multiple partition keys) the matching data rows if needed. As you can see, the bigger the size of the multiget, the slower the performance of the query because this will hit a lot of nodes to complete the fetch. You can optimize this query by storing the required data within the row, but denormalizing the dataset might be inflexible in some cases where the dataset changes more frequently. Choose the right model that fits your application.

    If most of your writes are updating the same key (hotspots on a key), all of the data can end up in a single node; to avoid these situations, you might want to consider partitioning the row key with some logic. For example, add a bucket or partition name to the end of the row key, or some sort of time-shading suffix:

```
Json Example: user_name : {
    1000: <value>,
```

```
      999: <value>,
      998: <value>, …
}
```

- **Key-value data store example**: Cassandra fits well as a key-value store too; the way to structure the data in Cassandra is a little different from any other key-value store. For example, if you want to store the query result in memcached, you would store the key as the query and the value as a blob of the result; whereas in Cassandra, you would store the key as the query but the value to be a pair of attribute and value. Cassandra provides richer data modeling abilities than a traditional key-value store. As a user you might need to choose what it should look like:

  ```
  Example: Query: {Column name:Value }
  ```

- **Distributed counters example**: Concurrent counters are as simple as incrementing the existing value by x; but in a distributed environment where network partitions and failures are normal operations, handling increments is a hard problem to solve. A timestamp is not enough to handle the failure cases. To overcome it, Cassandra handles the counters slightly differently than the normal writes. It is worth noticing that counters in Cassandra perform a read before write, which will cause additional load on Cassandra nodes.

Another gotcha is that counters in Cassandra are not idempotent, for example, the client writes to Cassandra, but before the client hears back from the server, there is an exception, and hence the client retries; this might cause overcounting. There is no way for the server to know if this count is a new count or an old one. There are some tickets open to rework counters to reduce this exposure.

Let's now talk about how you would use counters in Cassandra. Let's start by asking some questions:

- Do we need to handle the multi-DC situations? If yes, use it.
- Is accuracy really important? If yes, is it alright to overcount when there are exceptions?
- Is special handling of retries possible from the client? If yes, use it.
- Are there really any alternatives? If no, use it.

If the preceding questions don't fit your use case, don't use the counters. Modeling counters is similar to that of normal column families. You would structure the data as shown in the following line of code:

```
Map<byte[], SortedMap<byte[], Long>>
```

Let's look at an example of counting the number of retweets of a tweet.

Creating a counter column family

The following is how we create a counter column family:

```
CREATE TABLE tweets (
tweet_id varchar PRIMARY KEY,
application_id counter
) WITH comment='tweet Stats';
```

Tweet data structure

The data structure for the tweet example can be represented as follows, the key being the tweet_id column and the column name is application_id:

```
tweet_id :    {
        application_id:Count,
...
      }
Query:
SELECT * FROM users WHERE tweet_id= 'joedoe';
```

Secondary index examples

Cassandra supports secondary indexes. A secondary index in Cassandra is similar to that of RDBMS, but is different in functionality as it is a simple reverse/inverted index. Cassandra's built-in secondary indexes are a best fit for use cases when the values have high cardinality. In other words, the values occur more often and are less unique. Another option is to look at the collections data structure or wide-row functionality in Cassandra to fit some of these use cases.

Now the gotcha's:

- For every unique value, a row is created to hold the related keys which the node holds, and hence a secondary index query needs to scan the whole cluster

- A secondary index in most databases will do a read before write to update the index values, but since Version 1.2, Cassandra no longer performs read before write

Let's examine the Twitter profile example, where the user needs to look up all the users who are in CA. In the real case, you might want to add relevance to the mix, but let's consider this case.

Creating a secondary index table

The query to create a secondary index table is as follows:

```
CREATE TABLE users (
user_id uuid,
first_name text,
last_name text,
state text,
  PRIMARY KEY (user_id)
 );
```

Internal data structure

The following is how an internal data structure looks:

```
Joe.doe:    {
        first_name: Joe,
        last_name: Doe,
        state: CA
    }
```

Indexed column family

The following is how an indexed column looks:

```
CA:    {
        state: Joe_doe,
        ...
    }
```

Creating an index

The structure to create an index is as follows:

```
CREATE INDEX users_state
  ON users (state);
```

The query to get all the users in the state is as follows:

```
SELECT * FROM users WHERE state = 'CA';
```

Summary

In this chapter, we learned about CQL as a mechanism to insert and retrieve data out of Cassandra. We also talked about how a secondary index will help in searching for the data when the primary key is not known. In the following chapters, we will talk a little more on the internals of Cassandra.

In the next chapter, we will cover more on performance tuning based on the application's data model and usage. Having a better understanding of how the system works will help us understand Cassandra's performance profiles.

4
Administration and Large Deployments

In this chapter, we will talk about the basic administrative tasks and tools to manage data and its consistency.

There are three features in Cassandra that can make data consistent, and they are as follows:

- Hinted handoff
- Manual repair
- Read repair

Hinted handoff is the process in which if the write is not successful on a node or the node is not able to complete the writes in time, a hint is stored in the coordinator to be replayed at a later point in time when the node is back online.

The downside of this approach is that a node that has been down for a long time comes back online; all the nodes will start to replay hints in order to make the node consistent. These processes can eventually overwhelm the node with hint replay mutations. To avoid this situation, Cassandra replays are throttled by replaying a configured amount of bytes at a time and waiting for the mutations to respond; refer to `hinted_handoff_throttle_in_kb` to tune this number.

To avoid storing hints for a node that is down for a long time, Cassandra allows a configurable time window after which the hints are not stored (refer to `max_hint_window_in_ms` in `Cassandra.yaml`). For larger deployments, it might work fine, but might not work that well for a multi-DC setup. Hints replay will be naturally throttled by the latency (more than 40 milliseconds) that will cause the hints to replay forever. We can rectify this by increasing the number of threads involved in the replay (refer `max_hints_delivery_threads` in `Cassandra.yaml`).

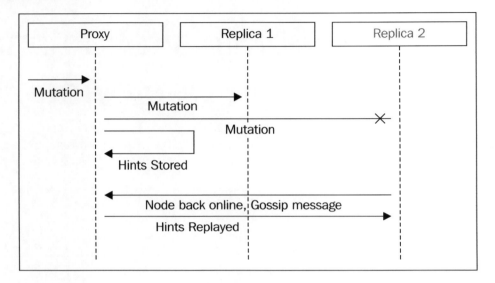

Manual repair

To make the data consistent, as administrators, we should repair the nodes at regular intervals. Manual repair or node tool repair is one such way to force Cassandra's nodes to sync the data. Manual repair or anti-entropy repair contains three stages:

1. Validation compaction
2. Exchange of Merkle trees and finding the inconsistencies
3. Streaming

The repair process starts by the user executing the command on a node to repair the given range of data. The node in turn sends a request for validation compaction to all the replicas of the range then waits for the compaction to complete. Validation compaction is the stage in which the node constructs a Merkle tree of the data. Once the Merkle tree is constructed as a result of the validation compaction, the nodes send back the Merkle tress to the coordinating node.

The coordinating node now has all the trees of the entire node for comparison to identify data inconsistencies. Once the inconsistencies are identified, the node sends the ranges to the node to stream inconsistent data back and forth.

It is worth noting that the coordinator doesn't know which data is right, and hence Cassandra exchanges the data both ways and lets the compaction fix the inconsistency. The basic idea of manual repair is to be network-friendly. We don't want to move all the data to be streamed and compared (which will be lot more expensive). In other words, `repair` is very similar to Linux `rsync`. If a file is found to be inconsistent, the whole file is sent across. Streaming is the phase of repair where the inconsistent data file chunks are exchanged between the nodes.

There are many cases where the repairs can be too expensive. For example, the depth of comparison for inconsistency is not tunable and large, thereby saving runtime memory usage. This can cause a bigger range to be streamed every time there is a small inconsistency in the range of interest, and this is followed by a huge compaction in the receiving end. To avoid this, Cassandra supports incremental repair that can be used to make the whole repair process predictable and quicker to complete.

The following command is for the node tool repair:

```
nodetool repair -pr
nodetool repair -st <Start token> -et <end token>
```

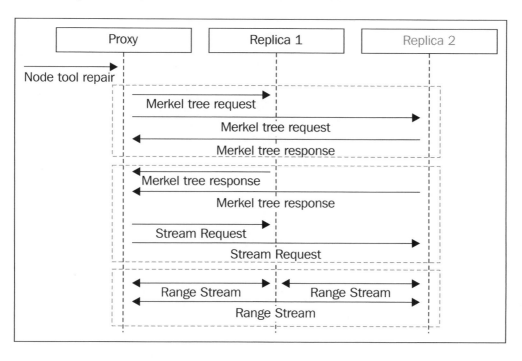

Bootstrapping

Bootstrapping is the process in which a newly-joining node gets the required data from the neighbors in the ring, so it can join the ring with the required data. Typically, a bootstrapping node joins the ring without any state or token and understands the ring structure after starting the gossip with the seed nodes; the second step is to choose a token to bootstrap.

During the bootstrap, the bootstrapping node will receive writes for the range that it will be responsible for after the bootstrap is completed. This additional write is done to ensure that the node doesn't miss any new data during the bootstrap from the point when we requested the streaming to the point at which the node comes online.

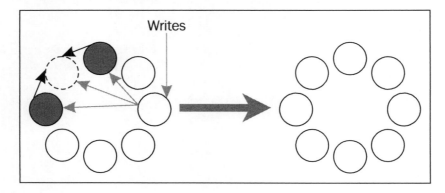

Bootstrapping works well when adding nodes to the ring, but there are cases when a node fails and we have to replace the node that has gone down as we know for sure that it is not going to come back online. Cassandra supports replacing a node with a different node or the same node with no data. To replace a node with the existing token, make sure that there is no data file in the directory and enable the system property during startup. The `-Dreplace_tokens=<tokens>` command tells Cassandra to bootstrap and replace the tokens.

With the replacement of tokens, the rest of the cluster will think the node is down but will start streaming data without sending read requests. In addition, the rest of the cluster is also storing hints for this node.

 You need to have the hint window open long enough to hold the data.

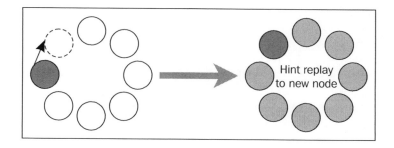

Once the node is up and running after streaming the required data, the hints are replayed into the node by the rest of the cluster.

Vnodes

Prior to vnodes, Cassandra nodes were assigned a single token for a node, causing the node to hold a large part of the contagious ranges. This will cause the bootstrapping node to wait for just two nodes that are next to each other. This creates an artificial limit on how fast the node can be bootstrapped, since streaming is throttled via Cassandra.yaml.

If instead of using vnodes we have a randomized token spread throughout the entire cluster, we still need to transfer the same amount of data. But now, there are more nodes streaming smaller. This allows us to rebuild the node faster than our single token per node scheme.

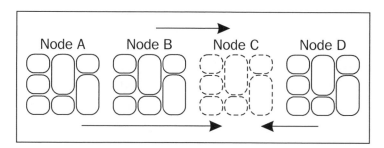

As you can see, the token ranges chosen are of random sizes, providing us with a better distribution of the ranges. Enabling vnodes is as simple as setting num_tokens in Cassandra.yaml, which will do the trick.

 With vnodes, you don't need to use replace_token; it is easy to remove a node and add a node.

Node tool commands

By now, you should have already understood the need for some of these administrative tasks to be performed on the Cassandra clusters. Now it's time to dive into the tools that will allow you to perform those actions on the cluster. Node tool is a command-line interface to run administrative commands on a Cassandra node, and sometimes on the whole cluster. The Node tool command-line tool can be found in the `bin` directory of the Cassandra distribution directory. The `nodetool --help` command shown previously is self-explanatory. In this chapter, we will focus on few of them. Cassandra exposes all the administrative functionalities via JMX, so users can also programmatically invoke those commands.

Cfhistograms

Cfhistograms displays statistics on the read and write latency. A histogram has buckets, and every time a request matches a bucket, it will be updated. You might want to read the columns that you are interested in and match it with the offset column.

 Every time Cfhistograms is queried, the data in it is reset. Hence, you might want to query at regular intervals to construct the histogram.

Cleanup

Cleanup starts a cleanup compaction of keys no longer belonging to this node. It is advised to run cleanup compaction after a node is moved or removed to clean up the data for which the node is not responsible for.

Decommission

Decommission executed on a node will make the node decommission itself (after streaming all the data to its neighbors in the ring).

Drain

Drain advises the node to flush all memtables and stop accepting writes. Read operations will continue to work:

- `move <token>`: This moves a node to the given token. The command simulates decommission and bootstrap to a new token.

- `netstats <host>`: This displays network information about streaming, triggered by the bootstrap, repair, move, or decommission commands. The pending SSTables to be streamed and its progress are good things to monitor to identify hangs.

- `rebuild <dc_name>`: This rebuilds data from some other datacenter. When rebuilding the data, this command is useful while bootstrapping a new datacenter or connecting a new datacenter to the existing ring.

- `removenode`: This removes a node by using a host ID from the ring. This command will be executed when the node is not a part of the cluster and for some reason you want to remove this node, as it will not join the cluster in the same token range. You can also force the removal by using the `removenode` force.

- `scrub`: This rebuilds SSTables on a node for the named column family and snapshot data files before rebuilding as a backup. While scrub rebuilds SSTables, Cassandra discards data that is corrupted.

- `setcompactionthroughput`: Use this if you want to override the `compaction_throughput_mb_per_sec` setting from YAML. Set the value to `0` to disable a compaction throttle without restarting the node.

- `setstreamthroughput`: Use this if you want to override the `stream_throughput_outbound_megabits_per_sec` setting in YAML. This sets the maximum streaming throughput in the system in megabytes per second (set `0` to disable throttling) without restarting the node.

- `snapshot`: This takes an online snapshot backup of Cassandra's data. Users can set up cron jobs to take an automatic snapshot of Cassandra and automatically move the data out of the node. `Snapshot` is a really a lightweight operation in Cassandra. All it does is create hard links to the existing stables. Cassandra creates hard links inside Cassandra's `data` folder in the `snapshot` directory. You can also enable incremental backup in YAML. Snapshot backup can be restored to the `data` folders and restarting the node can bring the data back to the node. Optionally, you can also apply the incremental backup to the restored snapshot.

> If snapshots are not cleared or if the files are not removed from the `snapshot` directory, the node might eventually run out of memory. Before `snapshot`, the node is flushed.

- **Stop**: This stops operations such as `compaction`, `validation`, `cleanup`, `scrub`, and `index build`. This allows you to stop a compaction that has a negative impact on the performance of a node. After the compaction stops, Cassandra continues with the rest in the queue.

- **Tpstats**: This displays the number of active, pending, and completed tasks for the thread pools that Cassandra uses for its operations. A high number of pending tasks for any pool can indicate performance problems.

Monitoring tools

Monitoring Cassandra is one of the most important tasks as an administrator. Cassandra provides a lot of useful metrics to help us understand a problem way ahead of time. For example, if your compaction time is constantly increasing, the SSTable size is increasing in exponential sizes, or if memory usage is increasing without the GC cleaning most of it, it is important for us to know about it and address the same.

DataStax OpsCenter

DataStax OpsCenter simplifies Cassandra monitoring and the management of Cassandra clustering the data infrastructure. OpsCenter is included as part of the DataStax distribution. It allows administrators, architects, and developers to manage, monitor, and control even the most complex database workloads with point-and-click ease from a centralized web browser. More details can be found at `http://www.datastax.com/`.

OpsCenter provides a good UI functionality of the `nodetool` command and collects the JMX monitoring data, and plots the data in intuitive ways. OpsCenter has both a community and enterprise version.

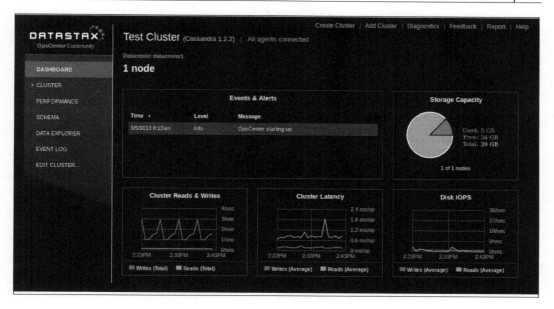

Basic JMX monitoring

Cassandra supports basic JMX monitoring, and hence there are varieties of tools that can be used to monitor the cluster. A simple Google search will reveal a variety of JMX monitoring tools; depending on your preferences, you might be able to find one you like. Hence, Cassandra can fit very well into any organization's enterprise monitoring solutions.

The node tool `netstats`, `compactionstats`, `tpstats`, and `cfstats` commands are invaluable in troubleshooting an issue. For example, if you have a spike in the latency, the first thing to refer is `cfstats` or the column family JMX metrics collected over a period of time. Basically, you might want to start by reasoning what changed causing this spike:

- Is there a spike in the number of requests per second?
- Are all the nodes up?
- Are there a lot of hints replayed?
- Is there a hot key? (check the average column size per column family)
- Is compaction spiking causing additional latencies?
- Are there a lot of threads pending for IO operation? Flush threads pending, read threads pending, and so on (if so, check IO wait time for disk issues)
- Is there a repair in progress? Is there a streaming in progress?

As you can see, monitoring the plain JMX can answer a lot of questions about what is happening in the cluster. By now, as an administrator, you might know the value of the historical data of the cluster; it will allow you to provision and also reason the clusters performance:

- **Push versus Pull**: Starting from Version 1.2, Cassandra switched to Yammer metrics, which provides a good extension to ganglia (http://ganglia.sourceforge.net). With little bit of development, you can customize the way the data is pushed to the monitoring system of your choice. Most of the JMX monitoring systems available in the market use JMX polling to report the data back to the monitoring service.

- **JConsole and VisualVM**: Java comes with Visual VM and JConsole which support basic Mbean monitoring and visualization of the basic metrics. VisualVM also helps in tuning the JVM. For example, in the following screenshot, note that the GC activity creates a saw-tooth pattern. If it doesn't, you might want to look into memory leaks.

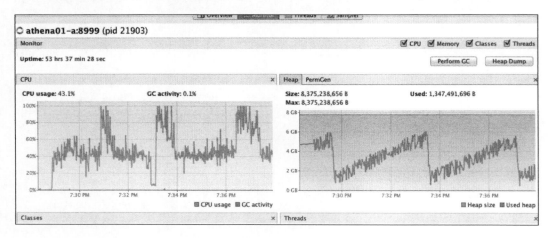

- **Mx4J**: Mx4j is a Java JAR file. If placed in the classpath, it will report the metrics via HTTP and in turn be parsed via the curl command, or can also be parsed by a complex SOAP parser.

Domain: org.apache.cassandra.db

```
org.apache.cassandra.db:type=BatchlogManager
org.apache.cassandra.db:type=BlacklistedDirectories
org.apache.cassandra.db:type=Caches
org.apache.cassandra.db:type=ColumnFamilies,keyspace=Keyspace1,columnfamily=Counter1
org.apache.cassandra.db:type=ColumnFamilies,keyspace=Keyspace1,columnfamily=Counter3
org.apache.cassandra.db:type=ColumnFamilies,keyspace=Keyspace1,columnfamily=Standard1
org.apache.cassandra.db:type=ColumnFamilies,keyspace=Keyspace1,columnfamily=Super1
org.apache.cassandra.db:type=ColumnFamilies,keyspace=Keyspace1,columnfamily=SuperCounter1
org.apache.cassandra.db:type=ColumnFamilies,keyspace=byte_array_store,columnfamily=byte_array_store
org.apache.cassandra.db:type=ColumnFamilies,keyspace=events,columnfamily=store
org.apache.cassandra.db:type=ColumnFamilies,keyspace=features,columnfamily=Comments
org.apache.cassandra.db:type=ColumnFamilies,keyspace=features,columnfamily=aggregate
org.apache.cassandra.db:type=ColumnFamilies,keyspace=features,columnfamily=aggregate2
org.apache.cassandra.db:type=ColumnFamilies,keyspace=features,columnfamily=lookup
org.apache.cassandra.db:type=ColumnFamilies,keyspace=features,columnfamily=lookup2
org.apache.cassandra.db:type=ColumnFamilies,keyspace=system,columnfamily=HintsColumnFamily
org.apache.cassandra.db:type=ColumnFamilies,keyspace=system,columnfamily=IndexInfo
org.apache.cassandra.db:type=ColumnFamilies,keyspace=system,columnfamily=LocationInfo
org.apache.cassandra.db:type=ColumnFamilies,keyspace=system,columnfamily=Migrations
org.apache.cassandra.db:type=ColumnFamilies,keyspace=system,columnfamily=NodeIdInfo
org.apache.cassandra.db:type=ColumnFamilies,keyspace=system,columnfamily=Schema
org.apache.cassandra.db:type=ColumnFamilies,keyspace=system,columnfamily=batchlog
org.apache.cassandra.db:type=ColumnFamilies,keyspace=system,columnfamily=hints
org.apache.cassandra.db:type=ColumnFamilies,keyspace=system,columnfamily=local
org.apache.cassandra.db:type=ColumnFamilies,keyspace=system,columnfamily=peer_events
org.apache.cassandra.db:type=ColumnFamilies,keyspace=system,columnfamily=peers
org.apache.cassandra.db:type=ColumnFamilies,keyspace=system,columnfamily=range_xfers
org.apache.cassandra.db:type=ColumnFamilies,keyspace=system,columnfamily=schema_columnfamilies
org.apache.cassandra.db:type=ColumnFamilies,keyspace=system,columnfamily=schema_columns
org.apache.cassandra.db:type=ColumnFamilies,keyspace=system,columnfamily=schema_keyspaces
org.apache.cassandra.db:type=ColumnFamilies,keyspace=system_auth,columnfamily=users
org.apache.cassandra.db:type=ColumnFamilies,keyspace=system_traces,columnfamily=events
org.apache.cassandra.db:type=ColumnFamilies,keyspace=system_traces,columnfamily=sessions
```

Summary

In addition to JMX monitoring, it is really important for us to monitor system metrics such as IO wait, network bandwidth utilization, swap usage (if any, it is recommended to disable swap on Linux), load average, CPU usage, and page in and page out activities. You can also monitor these metrics using `dstat`, `iotop`, `htop`, and various other tools available for the OS. Most monitoring tools will run a monitoring agent on the nodes to collect and aggregate the data for later use.

In the later chapters we will cover how to tune for better performance based on these metrics.

Performance Tuning

5

Cassandra's performance can be affected by any of the following:

- I/O contention between reads and writes (pending compactions)
- Frequent GC pauses
- Disk failures
- Network misbehaviors
- Threads and CPU contention

I would like to emphasize that a dead node in production is better than a slow node, and it is increasingly hard to identify a slow node that is to be replaced. This chapter will try to explain the tools and methods that can be implemented to identify the performance bottlenecks and take action on them or to tune your cluster for them.

In the past few releases, Cassandra has taken a lot of steps to simplify troubleshooting performance problems as well as improve the tooling around it. The Cassandra community has increasingly added features and improvements that can allow mixed workloads without any problems. Performance tuning depends on the types of operations your cluster performs most frequently and the type of queries performed on the data. I would like to talk about the various tools available in Linux before we start talking about performance tuning in Cassandra. It is really important as an administrator to constantly monitor system performance rather than monitor it only while tuning.

Let's start the chapter with an introduction to the following tools; these metrics can also be easily monitored regularly using most of the basic monitoring solutions. While reading this section, it will be nice if you could run the commands/tools on a test machine to get the context.

vmstat

vmstat reports information about processes, memory, paging, block I/O, traps, and CPU activity. When executed for the first time, the first line of the program returns the average since the machine was booted. This information is usually useless, hence it is recommended to run it in a loop (for example, the command vmstat 6 9 prints the metrics 9 times in an interval of 6 seconds).

```
-bash-4.1$ vmstat 6 9
procs -----------memory---------- ---swap-- -----io---- --system-- -----cpu-----
 r  b   swpd   free    buff   cache   si   so    bi    bo    in    cs us sy id wa st
14  0  14384 4052836 434496 108022624    0    0     4    16     2     1  3  0 97  0  0
 2  0  14384 3959608 434500 108114112    0    0     8    87 84350 78587 51  2 47  0  0
 1  0  14384 4014596 434504 108058208    0    0   133 36184 47124 72685 13  1 85  0  0
 0  0  14384 4012844 434504 108057568    0    0     0    73 18641 15960  2  0 97  0  0
12  0   4140 3784032 434504 108094800 1717    0  1719   169 141398 144802 35  4 60  1  0
 2  0   4140 3801072 434508 108072016    0    0  1918 24963 37211 35413 10  1 89  0  0
28  0   4140 3705612 434508 108119056    0    0  5879  5258 34306 45398  9  1 90  0  0
13  0   4140 3657568 434516 108194544    0    0  5235  6671 126671 127267 37  3 59  0  0
20  0   4140 3590908 434524 108245568    0    0  2990  5311 81787 98515 27  2 70  0  0
```

The elements of the preceding command line are elaborated in the following table:

Column	Description
buff	Amount of memory used as buffer
cache	Amount of memory used as cache
bi/bo	Blocks received and sent to the disks
in/cs	Interrupts and context switches
us	Time spent running non-kernel code
sy, st	Time spent running kernel code and the time stolen by a virtual machine

iostat

This reports CPU and input/output statistics for devices. The iostat command is used for monitoring system input/output device loading by observing the time the devices are active in relation to their average transfer rates. The first report generated by the iostat command provides statistics concerning the time since the system was booted and hence is mostly useless. It is recommended to run it in a loop (using a parameter; for example, iostats -mx 6 9). On multiprocessor systems, CPU statistics are calculated system-wide as averages, and are shown as follows:

```
-bash-4.1$ iostat -mx 6 9
Linux 2.6.39-200.24.1.el6uek.x86_64 (athena06-a.apple.com)     10/31/2013
    _x86_64_      (32 CPU)

avg-cpu:  %user   %nice %system %iowait  %steal   %idle
          2.82    0.08    0.35    0.03    0.00   96.73

Device:           tps   Blk_read/s   Blk_wrtn/s   Blk_read   Blk_wrtn

avg-cpu:  %user   %nice %system %iowait  %steal   %idle
         34.15    0.01    2.41    0.16    0.00   63.28

Device:           tps   Blk_read/s   Blk_wrtn/s   Blk_read   Blk_wrtn

avg-cpu:  %user   %nice %system %iowait  %steal   %idle
         25.93    0.00    2.40    0.19    0.00   71.48
```

The elements of the preceding command line are elaborated in the following table:

Column	Description
tps	Number of transfers (I/O requests) per second for the device.
Blk_read/s	Blocks read per second.
Blk_wrtn/s	Blocks written per second.
Blk_read	Total blocks read.
Blk_wrtn	Total blocks written.
r/s, w/s	The number of read and write requests that were issued to the device per second.
await	The average time (in milliseconds) for I/O requests issued to the device to be served. This includes the time spent by the requests in queue and the time spent servicing them.

dstat

Finally, I would like to introduce you to dstat. It is a versatile replacement for vmstat, iostat, and ifstat. dstat overcomes some of the limitations of these commands and adds some extra features. It allows you to view all of your system resources instantly; it also cleverly gives you the most detailed information in columns and clearly indicates in what magnitude and unit the output is displayed. dstat is unique in letting you aggregate block device throughput for a certain diskset or network bandwidth for a group of interfaces. And, its output is mostly self-explanatory.

Garbage collection

Garbage collection is the process of removing dead objects from the heap, allowing new Java objects to be created in the JVM heap memory. Garbage collection helps the JVM reclaim heap space/memory allocated by the threads processing read and write operations that are no longer needed. This process is performed by a set of threads that constantly clears and compacts memory for efficient memory management. Even though there are a lot of hurdles to overcome before getting the right GC settings, any decent-sized application written in any language has to deal with memory fragmentation. JVM's GC is implemented in a generic, well-understood manner, and has enough parameters to tune and make it efficient for high-performance applications.

Most new users are tempted to increase the heap size, thinking it will improve performance. In most cases, increasing the Java heap size is actually a bad idea, because JVM then has more memory to manage and has to frequently compact the memory locations, in turn increasing the CPU usage and causing larger stop-the-world GC pauses. In some cases, where the object allocation is slower, JVM doesn't have any incentive to keep compacting its heap, as there is more space available. GC will try to wait till it is $x\%$ full, but sometimes that might get too late to be able to catch up, in turn causing longer pauses/stop-the-world scenarios.

Cassandra also logs GC log information about garbage collection whenever garbage collection takes longer than 200 ms. If you start to notice a log message about the garbage collections occurring frequently and taking a moderate length of time to complete, it is a good indication that we need to start tuning the garbage collectors.

It is better to give most of the system's memory to the OS than allocating it to the JVM. Most operating systems maintain the page cache for frequently accessed data and are very good at keeping it in memory. In addition, we have to give enough memory to the JVM to cache its object creation. It is also worth noting that Cassandra also uses off-heap memory (more than what is configured in `cassandra-env.sh`) to store some of the data structures (such as `offheap`, `bloomfilter`, `indexsummaries`, and so on). Currently, Cassandra ships with **Concurrent Mark Sweep (CMS)** collector settings for its old-generation garbage collection, but this may change in future releases. We will cover CMS more than any other collector in this chapter.

CMS GC behavior looks similar to the following diagram; the objects are first created in **Eden** and eventually promoted to survivor spaces. During every promotion, the JVM will expire the objects that are not needed. Eventually, the objects that survive will be promoted to the old generation. If the threads try to create any object larger than the space available for the generations allocated, it will be created directly in **Old Generation**:

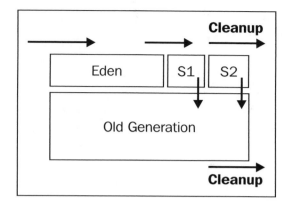

Enabling GC logging

It is strongly recommended to enable GC logging during performance tests; most users enable GC logging in production to better understand GC behavior when needed (GC logging has a very low overhead). Enabling the following in the `cassandra-env.sh` file can enable the required GC logging:

```
JVM_OPTS="$JVM_OPTS -XX:+PrintGCDetails"
JVM_OPTS="$JVM_OPTS -XX:+PrintGCDateStamps"
JVM_OPTS="$JVM_OPTS -XX:+PrintHeapAtGC"
JVM_OPTS="$JVM_OPTS -XX:+PrintTenuringDistribution"
JVM_OPTS="$JVM_OPTS -XX:+PrintGCApplicationStoppedTime"
JVM_OPTS="$JVM_OPTS -XX:+PrintPromotionFailure"
JVM_OPTS="$JVM_OPTS -XX:PrintFLSStatistics=1"
JVM_OPTS="$JVM_OPTS -Xloggc:/var/log/cassandra/gc-`date +%s`.log"
```

Understanding GCLogs

You can get more detailed information on how to interpret CMS by looking at all the numerous articles online; I recommend you visit `https://blogs.oracle.com/poonam/entry/understanding_cms_gc_logs`.

Let's start by examining some examples with explanations:

```
[GC 39.910: [ParNew: 261760K->0K(261952K), 0.2314667 secs]
262017K->26386K(1048384K), 0.2318679 secs]
```

The young generation collection's capacity was 261952K, and after the collection, its occupancy drops down to 0:

```
[GC40.704: [Rescan (parallel) , 0.1790103 secs]40.883: [weak
refs processing, 0.0100966 secs] [1 CMS-remark: 26386K(786432K)]
52644K(1048384K), 0.1897792 secs]
```

Stop-the-world GC

Stop-the-world GC is unavoidable most of the times, but we can make sure the full GC is not a major load on the system and can complete within a few milliseconds. A dynamic snitch usually tries to redirect traffic when it notices a non-responsive node. Consider the following commands:

```
[GC 197.976: [ParNew: 260872K->260872K(261952K), 0.0000688
secs]197.976: [CMS197.981: [CMS-concurrent-sweep: 0.516/0.531 secs]
(concurrent mode failure): 402978K->248977K(786432K), 2.3728734 secs]
663850K->248977K(1048384K), 2.3733725 secs]
```

This shows that a `ParNew` collection started, but was not able to complete because there was not enough space in the CMS generation to promote the object to the surviving young generation. Due to this, the concurrent mode of CMS was interrupted and a full GC was performed. Usually, when you see this, you might want to look if the Cassandra query is pulling a lot of data into memory; to get around this problem, you might want to increase the new size, increase the tenured generation size, or initiate the CMS collection at lesser heap occupancy by setting `CMSInitiatingOccupancyFraction` to a lower value.

The jstat tool

The `jstat` tool displays performance statistics for an instrumented HotSpot JVM; complete information on the options available can be viewed at `http://docs.oracle.com/javase/6/docs/technotes/tools/share/jstat.html`. The `jstat` tool is useful for understanding GC promotions and how much memory is utilized overall as well as the frequency of full GC.

As described in the earlier section, these numbers represent the object sizes in each generation, and you can see the promotion activities using the following commands:

```
-bash-4.1$ jstat -gcnew -h3 100507 1000
 S0C      S1C      S0U      S1U    TT MTT  DSS      EC         EU        YGC      YGCT
209664.0 209664.0    0.0 146799.2  1    1 104832.0 1677824.0 479742.1  56858 1863.935
209664.0 209664.0 125248.4 146799.2  1    1 104832.0 1677824.0 1677824.0  56859 1863.935
209664.0 209664.0 134223.8    0.0  1    1 104832.0 1677824.0 1520115.8  56859 1864.011
 S0C      S1C      S0U      S1U    TT MTT  DSS      EC         EU        YGC      YGCT
209664.0 209664.0 74910.1    0.0  1    1 104832.0 1677824.0 1273184.1  56861 1864.159
209664.0 209664.0    0.0 127738.9  1    1 104832.0 1677824.0 747168.0  56862 1864.190
209664.0 209664.0 129530.6    0.0  1    1 104832.0 1677824.0 680480.4  56863 1864.272
```

The following commands show how the objects are allocated and promoted; for a healthy VM, most objects will expire during promotions:

```
$ jstat -gc -h5 18753 1000 100
```

```
-bash-4.1$ jstat -gc -h5 100507 1000 100
 S0C      S1C     S0U  S1U     EC        EU         OC        OU          PC       PU      YGC    YGCT    FGC  FGCT     GCT
209664.0 209664.0 0.0  56241.3 1677824.0 1615447.5 6291456.0 2862930.2 78468.0 46790.4  56944 1868.942 476  146.605 2015.547
209664.0 209664.0 129973.6 0.0  1677824.0 1359708.0 6291456.0 2862954.5 78468.0 46790.4  56945 1868.971 476  146.605 2015.576
209664.0 209664.0 0.0  105512.4 1677824.0 1330398.7 6291456.0 2955535.0 78468.0 46790.4  56946 1869.050 476  146.605 2015.655
209664.0 209664.0 127372.1 0.0  1677824.0 1158605.6 6291456.0 3026247.6 78468.0 46790.4  56947 1869.116 476  146.605 2015.721
209664.0 209664.0 0.0  115356.1 1677824.0 973885.4 6291456.0 3107422.7 78468.0 46790.4  56948 1869.186 476  146.605 2015.791
```

```
$ jstat -gcutil 18753 1000 100
```

```
-bash-4.1$ jstat -gcutil 100507 1000 100
  S0     S1      E      O      P      YGC     YGCT    FGC    FGCT     GCT
 0.00  55.08  56.17  75.00  59.63  57050 1874.754  479  146.788 2021.542
 0.00  37.36  34.80  77.65  59.63  57052 1874.889  479  146.788 2021.677
 0.00  37.36  96.32  76.32  59.63  57052 1874.889  480  146.896 2021.785
89.43   0.00  96.21  65.47  59.63  57054 1874.927  480  146.896 2021.823
 0.00  54.69  82.28  56.55  59.63  57054 1875.009  480  146.896 2021.905
53.41   0.00  64.12  46.76  59.63  57055 1875.066  480  146.896 2021.962
15.65   0.00  21.49  41.01  59.63  57057 1875.209  480  146.896 2022.105
 0.00  54.75  10.51  31.77  59.63  57058 1875.237  480  146.896 2022.133
 0.00  54.75  96.45  28.60  59.63  57058 1875.237  480  146.896 2022.133
57.16   0.00  64.51  30.01  59.63  57059 1875.317  480  146.896 2022.213
```

The jmap tool

The `jmap` tool can be used to understand the types of objects that are currently occupying the memory. If you detect a memory leak, it might be worth taking a heap dump for analysis. `jmap` can be used for a cheap analysis of memory utilization as shown in the following command line snippet:

```
$ jmap -histo:live 18753
```

```
num     #instances        #bytes  class name
----------------------------------------------
  1:      4227846     915402456  [B
  2:      5362720     257410560  java.nio.HeapByteBuffer
  3:      2282687      91307480  org.apache.cassandra.db.ExpiringColumn
  4:      1329819      63831312  edu.stanford.ppl.concurrent.SnapTreeMap$Node
  5:        13072      60628312  [S
  6:        25638      56004472  [I
  7:       465526      53002784  [Ljava.lang.Object;
  8:      2012519      48300456  java.lang.Double
  9:      1332783      42649056  java.util.concurrent.ConcurrentHashMap$HashEntry
 10:      1310670      41941440  com.googlecode.concurrentlinkedhashmap.ConcurrentLinkedHashMap$Node
 11:      1527898      36669552  org.apache.cassandra.db.DeletionInfo
 12:      1123907      35965024  java.util.ArrayList$Itr
```

The write surveillance mode

To get the right settings and tuning in place, it is very important for the users to test real-world scenarios. To achieve this, Cassandra supports the write survey mode without responding back to the read traffic. The user can change and tune write traffic, compaction, and compression strategies, and these can be compared with the rest of the cluster. Using the write survey mode, users can experiment with different strategies and benchmark compaction, write performance, and so on. They can also note all the differences in tuning without affecting production workload. It is worth noting that in the write survey mode, read traffic cannot be tested. To enable write survey mode, start a Cassandra node with `-Dcassandra.write_survey=true` appended to the start up script.

Tuning memtables

When Cassandra receives a write request, it is first written to the commit logs (queued to be written) and then to a sorted in-memory structure called a Memtable, maintained per column family. When a Memtable is full (configured size in-memory), it is flushed to disk as an SSTable. A memtable is basically a write-back cache of data rows that can be looked up with the help of a key; that is, unlike a write-through cache, writes are queued up in the Memtable.

Tuning memtable involves understanding your write requests; if you have a lot of writes to be performed (or a lot of compaction queued up), increasing the size in memory will reduce the amount of compaction since the SSTs will be bigger before they are flushed, and hence the compaction will be on bigger files rather than on several small ones. Memtable thresholds are configured per node using the Cassandra. yaml property, memtable_throughput_in_mb. You might also want to increase the sizes if you have a lot of updates to the same key in a short period of time. For leveled compaction, you can choose the level0 sizes such as sstable_size_in_mb while creating the column family (this can also be updated later); the default size is 4 MB.

memtable_flush_writers

The memtable_flush_writers setting controls the number of threads used for flush writers. If there are more than one disks involved in writing memtables to the files, it will help parallelizing the number of writers. If an SSD drive is used, users might see improvements in flush performance by increasing the number of threads. You might want to monitor the FlushWriter elements of tpstats to discover the bottlenecks, as shown in the following diagram:

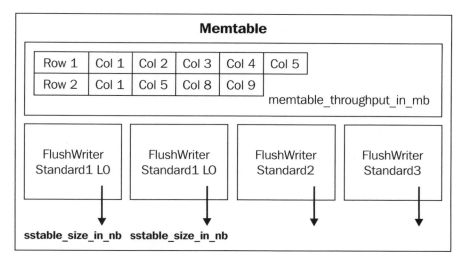

Compaction tuning

Compaction tuning starts by first selecting the right compaction strategy. Once we choose the right strategy for the workload, we can start tuning the parameters to make them more efficient.

SizeTieredCompactionStrategy

This strategy is best suited for column families with higher write-to-read ratios. This compaction strategy compacts similar sized SSTables into a bigger file when the similar sized SSTables' `min_compaction_threshold` is reached. The maximum number of files that are involved in a compaction can be controlled by `max_compaction_threshold`.

By setting `min_compaction_threshold` to a larger value, we can force Cassandra to delay compacting multiple small files, reducing the amount of IO required for compaction. For example:

- 4 * 1 GB (Memtable limit) = 4 GB
- 4 * 4 GB (second phase) = 16 GB

Instead, if the value of `min_compaction_threshold` is 10:

- 10 * 1 GB = 10 GB
- 10 * 10 GB = 100 GB (second phase)

If you want to reduce `max_compaction_threshold`, or if you want the compaction to be faster by involving a smaller number of SSTables, it requires more disk IO to process it:

```
> UPDATE COLUMN FAMILY users WITH max_compaction_threshold = 20;
```

LeveledCompactionStrategy

The compaction strategy is based on Google's `leveldb` library with some significant modifications. This strategy is best suited for column families with higher read-to-write ratio workloads and/or column families with a lot of updates to existing rows. Any given primary key in leveled compaction is almost guaranteed to be in at most four SSTables, providing better read-performance benefits.

The downside of this strategy is that Cassandra will compact a lot more than the sized strategy. Consider the following example:

```
> UPDATE COLUMN FAMILY users WITH compaction_strategy=LeveledCompactionSt
rategy
AND compaction_strategy_options={sstable_size_in_mb: 10};
```

After choosing the right strategy for the use case, compaction can further be tuned using the following settings:

```
in_memory_compaction_limit_in_mb
compaction_throughput_mb_per_sec
concurrent_compactors
multithreaded_compaction
compaction_preheat_key_cache
```

Compression

Compression reduces the data and disk utilization of a Cassandra node, subsequently reducing the need for disk IO. When Cassandra needs to look for data, it looks up the in-memory chunk offset and unpacks the data chunk to get to the columns. Compression is a CPU-intensive process, hence you might want to keep an eye on CPU usage when compression is enabled. Both writes and reads are affected because of the chunks that have to be compressed and uncompressed. On the flip side, compression of data means less data to write and hence better performance.

```
CREATE TABLE Standard1 (
        id text,
        column_name text,
      PRIMARY KEY (id)
      )
        WITH compression = { 'sstable_compression' :
'DeflateCompressor', 'chunk_length_kb' = 64};
```

The bigger the compression chunk length, the better the compression ratio for the SSTables. The downside of bigger chunk size is that while decompressing/reading data from the disk, we have to uncompress a bigger chunk, which can be expensive both in terms of memory complexity and time complexity. Compression has a high CPU cost on reads, but is better on page caches (there are less blocks to cache). Most write-intensive applications will benefit from compression. It is worth noting that compression metadata has memory overhead on smaller compression chunk sizes.

NodeTool

NodeTool is the command-line tool that communicates via JMX to aggregate various metrics from the Cassandra node and then prints them on the console. It is worth noting that you can get all that information by monitoring the JMX from a generic monitoring tool. Since Version 1.2, Cassandra uses a metrics library that exposes a lot more information than previous releases (including 99 Percentile, 15 min, 5 min averages, and so on).

compactionstats

compactionstats provides compaction progress information of the node. With leveled compaction, it is normal to see a continuous compaction, but if the pending tasks are in the 100s and never catch up, this is something we need to worry about. Compaction is Cassandra's self-healing process to reduce the number of SSTables required to consult before responding to the read requests. This is demonstrated as follows:

```
-bash-4.1$ ./apache-cassandra-1.2.8/bin/nodetool compactionstats
pending tasks: 1
        compaction type     keyspace   column family      completed        total     unit   progress
            Compaction        events          store       88062223    595134249    bytes     14.80%
Active compaction remaining time :    0h00m03s
```

If you detect the compaction backing up, you can choose to disable compaction or increase compaction throttle and increase speed. But this tuning is a tradeoff between I/O available for read queries versus I/O available for compaction. Monitoring IOWait will reveal the I/O contentions, if any.

If you notice this behavior while using SSDs, the contention will be in-memory, and you might want to be careful before increasing in_memory_compaction_limit_in_mb; this might cause increased GC activity. This can be better tuned by increasing the value of HEAP_NEWSIZE, but care should be taken to avoid a huge generation, which can cause bigger GC pauses.

netstats

netstats shows the existing stream progress from and to the node, allowing users to monitor progress (and is not stuck due to network issues). Cassandra retries in case the connection breaks. Streaming happens when node bootstraps and data have to be streamed from various nodes. Streaming may also be performed when a node is repaired (when repair finds an inconsistency and decides to stream inconsistent data front and back). Cassandra retries if the connection for the file stream breaks.

tpstats

The following output shows the number of active and pending requests. If you notice a lot of pending requests backing up, it is a good indication to increase the number of concurrent_reads and concurrent_writes variables or increase the cluster size to handle the load as shown in the following command-line snippet:

```
-bash-4.1$ ./apache-cassandra-1.2.8/bin/nodetool tpstats
Pool Name                    Active   Pending   Completed   Blocked   All time blocked
ReadStage                         0         0   131912895         0                  0
RequestResponseStage              0         0    93711853         0                  0
MutationStage                     0         0    40848464         0                  0
ReadRepairStage                   0         0     4648545         0                  0
ReplicateOnWriteStage             0         0           0         0                  0
GossipStage                       0         0     2516463         0                  0
AntiEntropyStage                  0         0           0         0                  0
MigrationStage                    0         0          12         0                  0
MemtablePostFlusher               0         0        9907         0                  0
FlushWriter                       0         0         684         0                 19
MiscStage                         0         0           0         0                  0
commitlog_archiver                0         0           0         0                  0
InternalResponseStage             0         0           0         0                  0
HintedHandoff                     0         0          60         0                  0

Message type           Dropped
RANGE_SLICE                  0
READ_REPAIR               1891
BINARY                       0
READ                        16
MUTATION                 12507
_TRACE                       0
REQUEST_RESPONSE             9
```

The commands being dropped are another thing worth noticing. If you observe a lot of drops in any of the stages, it might be an indication of performance problems; it could also mean there are some random network issues. It is worth monitoring the logs to see if the nodes are more frequently flapping up and down during the operation. If that's true, it could be an indication of GC pauses or more serious problems. Hence, more investigation along the lines discussed in this chapter might be needed.

Cassandra's caches

The key_cache_size_in_mb and row_cache_size_in_mb settings in Cassandra. yaml allow a user to tune the caches in Cassandra. These settings are global settings for the whole node that allow users to specify the amount of memory that has to be reserved for key caches and row caches during startup. Because both caches take some part of the heap memory, it is important for us to tune it ahead of time. Row cache is off-heap by default, but the keys are stored in a heap, hence care should be taken while tuning it.

Cassandra's built-in key and row caches can provide very efficient data caching in some cases and can be abused in others. There are some cases where the production deployments have leveraged Cassandra's caching to replace distributed caches. Using caching in Cassandra can remove cache coherence problems and eliminate the need for additional logic that needs to be written in the client during reads and writes.

You can tune Cassandra's caches by enabling GC logging or watching GC activity via visual VM or any other tool during the load. If a read operation hits the row cache, the entire requested row is returned without disk seeks. If a row is not in the row cache but is present in the key cache, the key cache is used to find the exact position of the row in a particular SSTable. If a row is not in the key cache, the read operation will populate the key cache after accessing the row on disk, so subsequent reads of the row can benefit.

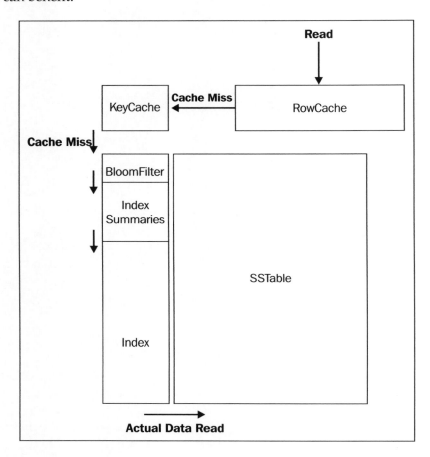

Consider the following line of code:

```
>UPDATE column family Standardl with caching=ALL;
```

The supported values are as follows:

- -All (enables the row cache and key cache)
- -KEYS_ONLY

Filesystem caches

It is really important to make sure Cassandra's heap is not swapped. To avoid swapping, Cassandra tries to lock the JVM heap in-memory, but Cassandra also uses off-heap structures; hence, it is important to monitor system swap activities to find any abnormalities. Most users disable swap to avoid this situation, which is completely normal.

Most modern operating systems try to optimize the reads and writes in the system by properly caching the filesystem blocks for faster access. It is completely normal to see 100 percent usage of the memory in cases where most of the memory location is cached. Memory has increasingly become cheaper, and the more memory available for caches, the faster the reads for frequently accessed blocks. It might not be helpful in cases where there are a lot of writes because disk throughput for cache and compaction becomes a bottleneck for those cases:

```
-bash-4.1$ free -m
              total       used       free     shared    buffers     cached
Mem:         129148     124807       4340          0        425     104352
-/+ buffers/cache:       20029     109118
Swap:          4095          4       4091
```

Separate drive for commit logs

A commit log in Cassandra is mostly sequential, and the flush will happen depending on the following settings. You might want to configure commit logs in a separate filesystem for them to be purely sequential. Because every write will try to write to the filesystem, it might be helpful to not mix the normal reads with the commit logs. An example of a commit log is as follows:

```
commitlog_sync: batch/periodic
```

Tuning the kernel for Cassandra

There are multiple kernel tuning parameters that can provide marginal performance gains in addition to the previously mentioned tunings. It is important to note that you might want to tune the following only after you are reasonably sure there aren't any bottlenecks and you are comfortable with the performance of the system as a whole. Talking about multiple operating systems is out of the scope of this book, hence we will just cover tuning in Linux.

```
# Limit of socket listen() backlog, known in user space as SOMAXCONN.
Defaults to 128.

net.core.somaxconn = 2048
```

```
# Sets the default and max OS receive and send buffer size for
connections
net.core.rmem_default = 131072
net.core.wmem_default = 131072
net.core.rmem_max = 8388608
net.core.wmem_max = 8388608
# Allow enough memory to actually allocate those massive buffers to a
socket.
net.ipv4.tcp_mem = 8388608 12582912 16777216
# Disable selective acks. This should be tested with and without. Given
the size of the buffer, it should not make a material difference.
net.ipv4.tcp_sack = 0
net.ipv4.tcp_dsack = 0
net.ipv4.tcp_fack =0
# Try to get rid of connections that are likely bad. Defaults are really
conservative.
net.ipv4.tcp_keepalive_time = 300
net.ipv4.tcp_keepalive_probes = 9
net.ipv4.tcp_keepalive_intvl = 30
net.ipv4.tcp_fin_timeout = 30
```

noop scheduler

It is clear that Cassandra will benefit from filesystem caches; the default Linux scheduler is cfs, which will try to cache the blocks depending on the previous metrics. But this might not be a good thing for Cassandra; we would like to cache the blocks that are required to be accessed. The noop scheduler does just that.

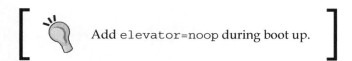

Add elevator=noop during boot up.

NUMA

Modern multi-core machines support the NUMA architecture, which basically caches the memory locations in threads. In Cassandra, it is hard to pin a memory location to threads, and hence we might want to disable NUMA on the system. Cassandra's startup script performs this by default, but it is important to install numactl on the system.

Other tuning parameters

The following are some other tuning parameters:

- **Increase Map Count**: Default `mmap` segments allowed by the VM are too small; this is most obvious while using leveled compaction, as shown in the following line of command:

```
$ sysctl -w vm.max_map_count=271360
```

- **Increase User limits**: Increases the allowed user limits so Cassandra can take advantage of the system resources available. This is illustrated as follows:

```
$ echo "
* soft nproc 100000
* hard nproc 100000
 * soft nofile 200000
 * hard nofile 200000
 root soft nofile 200000
 root hard nofile 200000
 * soft memlock unlimited
 * hard memlock unlimited
 root soft memlock unlimited
 root hard memlock unlimited
 * soft as unlimited
 * hard as unlimited
 root soft as unlimited
 root hard as unlimited
 " |sudo tee -a /etc/security/limits.conf
```

Dynamic snitch

A snitch determines which datacenters and racks are both written to and read from. Writes in Cassandra cannot be benefited from dynamic routing and cannot be optimized, as all the replicas need to receive the write requests. However, reads in Cassandra request a node for the actual data and the remaining replicas for `md5sum` (depending on the consistency level and read-repair chance configured).

Because it is one node that responds with the data and the rest will be serving `md5sum`, it is important for Cassandra to choose the right node so the read request can be served without dropping the request. To decrease the chance of failure, a dynamic snitch uses the disk unitizations of the nodes (gossiped information), latency seen by previous requests, and the `DC/RACK` information to make an informed decision. It might be worth noting that the recent release of Cassandra has a thread that constantly monitors the `IOWait` time on the nodes and reports its gossips to other nodes, which is used by Cassandra's dynamic snitch to route traffic.

By default, Cassandra chooses a different node if it is more than 10 percent worse than the other node. You can choose to increase or decrease this by changing the value of `dynamic_snitch_badness_threshold`. On the flip side, reducing the threshold will cause the reads to be redirected and cause them to be scattered too much, hence less stickiness.

The following diagram illustrates the idea of a dynamic snitch; the algorithm and the variables can change from release to release:

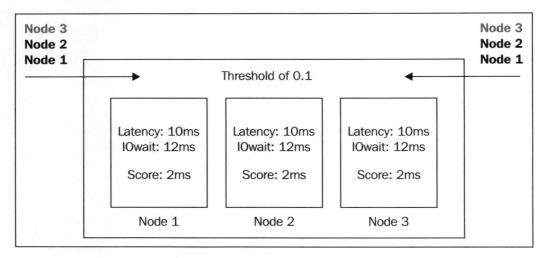

Configuring a Cassandra multiregion cluster

If you choose to go for the `vnode` option for your multi region Cassandra cluster, configuring the cluster is a no operation, as most buckets are randomly chosen within datacenters. `num_tokens` in `Cassandra.yaml` allows users to choose the number of partition/tokens per node (defaults to 256). To randomize the tokens and their mapping to the node, use the `shuffle` operation. The `shuffle` command shuffles the existing mapping of data partitions. It's important to note that the `shuffle` command is not required for new installations; rather, it basically moves the data around the cluster, rebalancing the cluster. `VNode` provides better distribution of data even when using byte-ordered partitions / order-preservative partitions (not recommended) in addition to providing us with faster rebuild and repair.

Repair is an administrative task that needs to be executed before the GC grace period is reached. If your repair takes a long time to complete in your cluster, try incremental repair to reduce the cost of repairs significantly.

If you choose to go for manual token assignment for some reason, it is important to make sure we have an even distribution of tokens. There are various scripts available online to calculate the tokens; pick one and make sure the ring shows an even distribution of tokens, as shown here:

```
-bash-4.1$ nodetool ring features

Datacenter: DC1
==========
Replicas: 2

Address         Rack        Status State   Load        Owns        Token
                                                                    -3757670089088982778
17.149.128.172  RAC6        Up     Normal  9.45 GB     11.11%      7173733806442603386
17.149.128.173  RAC6        Up     Normal  9.46 GB     11.11%      9223372036854775782
17.149.128.103  RAC4        Up     Normal  9.46 GB     11.11%      -4440882832559706910
17.149.128.170  RAC6        Up     Normal  9.46 GB     11.11%      3074457345618258594
17.149.128.171  RAC6        Up     Normal  9.5 GB      11.11%      5124095576030430990
17.149.128.102  RAC4        Up     Normal  9.46 GB     11.11%      -6490521062971879306
```

It is also important to note that we have the even distribution within a region; it is worth noting that any node in the same cluster cannot have any random token collusion. Users have to assign unique tokens to the nodes in the cluster. The network topology strategy is responsible for choosing the nodes in the cluster for a particular token. While configuring the NTS, the user needs to choose the number of replicas per DC, which is used during the node selection for both reads and writes.

As shown in the following diagram, when a read reaches the proxy, the proxy consults the NTS to find the nodes responsible for the token, for which the NTS responds back with a set of servers depending on the configured replication factor.

```
> CREATE KEYSPACE Keyspace1 WITH strategy_class =
'NetworkTopologyStrategy' AND strategy_options:DC1 = 2 AND strategy_
options:DC2 = 2;
```

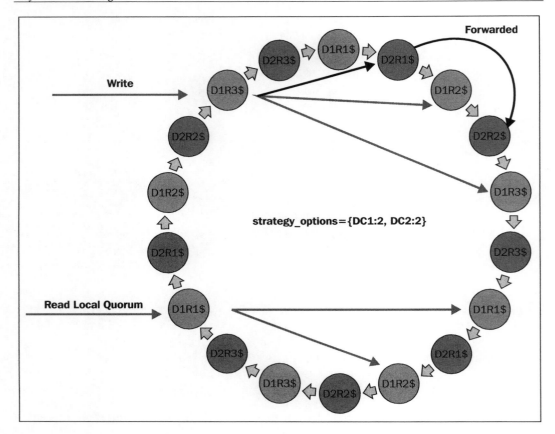

Reads are a little different; even though the proxy/coordinator receives all of the nodes in the cluster, it filters the nodes that should be skipped and sends the request based on the consistency level and read-repair requirements. The coordinator sends the data request for the first node and the rest will receive the checksum request.

Summary

GC tuning is not as straightforward a topic to be covered in detail in this chapter, but the tools discussed in this chapter will get you started. Recent releases of Cassandra have worked around the limitations of JVM to give a better experience to the user; for example, most of the data structures that consume large enough memory are moved out of the heap. Cassandra 2.0 is a big stride forward in moving almost all the big memory structures off-heap. In fact, most of the installation will not see any or will see very little memory pressure in most common use cases. In the later chapters, we will cover more about troubleshooting.

6
Analytics

It is important to understand that the analytics workload has characteristics that are different from normal reads and writes in Cassandra. In most cases, MapReduce jobs will have to read all or most of the data from Cassandra to run the analytics on it. Mixing the analytics workload and normal reads can cause poor performance on the regular read operations in the system. It will cause filesystem caches and other Cassandra caches to be trashed by the analytics workload. Hence, most users separate different workloads into different physical clusters, allowing the normal reads to be optimized by Cassandra.

It is important to note that this scenario is true not only for Cassandra, but for any other real-time data storage solution available in the market. There are a lot of analytics products the community supports; we will just cover a few of them in this chapter.

Hadoop integration

Cassandra out of the box supports Hadoop, Hive, and Pig to run analytics jobs on top of existing data in Cassandra.

Configuring Hadoop with Cassandra

There are multiple ways to configure Hadoop to work with Cassandra. It's best to run the Hadoop cluster on the same server as Cassandra nodes, which goes with the idea of moving the computation to the data than moving the data to the computation. Database vendors such as DataStax provide a Hadoop distribution, which allows us to run Hadoop on the same box as Cassandra, removing the need for HDFS storage for intermediate data and metadata for the jobs.

Alternatively, you want to have a separate server for your NameNode and JobTracker and install a Hadoop TaskTracker and DataNode on each of your Cassandra nodes, allowing the JobTracker to assign tasks to the Cassandra nodes that contain data for those tasks. You also need HDFS installed on different drives or nodes to allow Hadoop to use the distributed filesystem for copying dependency jars, static data, and intermediate results for job execution. This method allows users to run the analytics near the data instead of transferring the data across the network, but it is worth noting that this might be a bad idea in some cases since the resources are shared.

Users might also want to disable dynamic snitch while trying to do local reads on the analytics nodes; this will force Cassandra to read locally instead of fetching the data from the neighbors when the score fluctuates on the basis of latencies. You also need to configure `HADOOP_CLASSPATH` in `<hadoop_home>/conf/hadoop-env.sh` to include the Cassandra `lib` libraries.

Virtual datacenter

You can configure the virtual datacenter (just like a multi-DC setup, but it's not physically separated) in Cassandra to separate the analytics workload from the real-time web serving (or similar) traffic; before changing any of the settings, make sure that all the keyspaces use **NetworkToplogyStrategy** (**NTS**) and the required replication factor is already selected. (Please see the previous chapters for choosing the replication factor and configuring the keyspaces.)

Adding nodes to a different datacenter (virtual) is as simple as changing the property files if `PropertyFileSnitch` or `GossipingPropertyFileSnitch` are used. Configure the new nodes with the appropriate tokens or enable vnodes to start these nodes.

There are two ways to bootstrap the node with the data that the node is supposed to have. However, the recommended and faster way to bootstrap the new cluster is by enabling the replication factor of the new DC for the keyspaces that need to be replicated, and bootstrapping the nodes in the new cluster one-by-one. The other way to bootstrap the datacenter is to bootstrap the node without changing the replication factor and later repairing the nodes one-by-one to synchronize and replicate the old data to the new DC.

PropertyFileSnitch

`PropertyFileSnitch` helps Cassandra know about the location of the nodes by rack and datacenter. Snitch uses a defined description of the network information from the property file located in `<Cassandra_Home>/conf/cassandra-topology.properties`; or if you use the Debian package, look for the `/etc/cassandra` folder (same location as `Cassandra.yaml`).

As a user, it is your responsibility to configure the right IP address and server location, which are used by Cassandra to understand the network layout.

You can define your datacenter names to be any arbitrary names, but care should be taken to match the datacenter name provided here to the name provided while configuring the keyspace's strategy options (NetworkToplogyStategy).

Every node in the cluster should be described in the `cassandra-topology.properties` file and should be consistent with all the servers in the cluster.

The following is a sample configuration used to configure the analytics node:

```
10.0.0.10=DC1:RAC1
10.0.1.10=DC1:RAC2
10.0.2.10=DC1:RAC3
10.20.0.10=Analytics:RAC1
10.20.1.10= Analytics:RAC2
10.20.2.10= Analytics:RAC3
# default for unknown nodes
default=DC1:RAC1
# Native IPv6 is supported; however you must escape the colon in the IPv6
fe80\:0\:0\:0\:202\:b3ff\:fe1e\:8329=DC1:RAC1
```

GossipingPropertyFileSnitch

Since `PropertyFileSnitch`, Cassandra has implemented a new version that uses the gossiper (`GossipingPropertyFileSnitch`) to understand the location of the servers. It allows the user to define a local node's datacenter and rack, hence allowing gossip to propagate information to other nodes. It starts by updating `<cassandra_home>/conf/cassandra-rackdc.properties`.

For example, in DC1 DC, RAC1:

```
dc=DC1
rack=RAC1
```

For example, in Analytics DC, RAC1:

```
dc=Analytics
rack=RAC1
```

As you can see, the previous snitch is similar to Ec2Snitch, which uses the AWS APIs to understand the node's location. Ec2Snitch also supports `dc_suffix` for appending to the AWS provided name, so we can have the analytics workload coexist in the same Amazon DC. For example:

```
dc_suffix=-Analytics
```

DSE Hadoop

The advantage of using DSE Hadoop is that it is a fully integrated solution. There is no need for the NameNode and TaskTracker to be set up as separate installations. The installation is seamless and it is as simple as starting Cassandra using the following command:

```
# dse cassandra -t
```

The good part about this is that DSE implements HDFS APIs in Cassandra. Hence, there is no need to have a separate filesystem for the intermediates. This also simplifies the metadata store; while using Hive, it is an integrated solution. More detailed information can be found at http://www.datastax.com/.

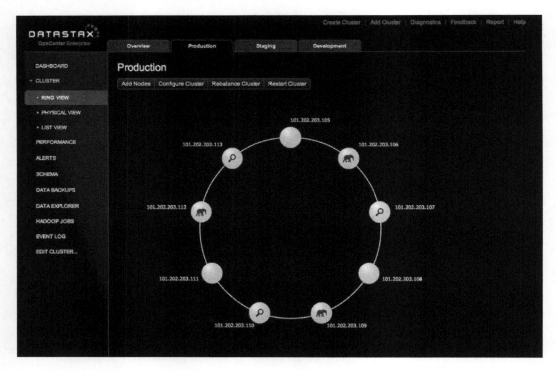

(Courtesy: www.datastax.com)

Acunu Analytics

Acunu Analytics allows users to maintain roll-up cubes, just like the OLAP databases on the market, as and when the data is ingested internally using Cassandra counters to hold the roll-ups in memory. In some ways, this is similar to a facet search in Solr. The data queried on these cubes reflects fresh data immediately in queries that are close to instantaneous because they typically only retrieve precalculated results. But this also means additional work during the updates. It is also important to note that Acunu performance characteristics will be completely different from what this book talks about and out of context to most of the optimizations mentioned in this book.

Reading data directly from Cassandra

This is based on the implementation of `ColumnFamilyInputFormat` and `ColumnFamilyRecordReader`, so that Hadoop MapReduce jobs can retrieve data from Cassandra. Cassandra rows, or row fragments (that is, pairs of keys and columns of SortedMap), are input to map tasks to process your jobs as specified by a SlicePredicate that describes which columns to fetch from each row. You can also install the TaskTracker and the DataNode on the same node as Cassandra and simulate the same performance as DSE can provide.

Analytics on backups

Sometimes, it is economical to not have multiple servers running all the time for analytics that run once in a while, and there may be alternative and cheaper storage mechanisms available. In this case, you can use the incremental backup that Cassandra supports to get data out of the server and read it to extract data out of the SSTables; if done right, this storage can also serve as your backup system.

There are multiple ways to achieve this; one well-known way is to convert the backups into JSON by using `sstable2json`, which converts the on-disk SSTable representation of a column family into a JSON formatted document. Its counterpart, `json2sstable`, does exactly the opposite. But we need to make sure we resolve the conflicts on the JSON documents that have been pulled out (which we are not going to cover in this chapter). Another elegant way is to write a generic program that can read the SSTable format.

File streaming

Writing output for the MapReduce task is a normal operation that will be done as part of the analytics. An example of it would be calculating an aggregate of the data for the past hour and writing it back to Cassandra to serve the data. Cassandra comes with a bulk uploader utility that can be used by Hadoop MapReduce to upload the output of the reduce job. The advantage of using the bulk uploader is that it doesn't use Thrift to write back to Cassandra, and hence it is faster.

It is important to understand what the sstableloader tool does; given a set of SSTable data files, it streams them to a live cluster. It does not simply copy the set of SSTables to every node, but transfers only the relevant part of the data to each node, thus conforming to the replication strategy of the cluster. But this requires the user to create the SSTables that need to be streamed. One good thing about MapReduce is that you can sort the keys before writing the SSTables, and hence you may be able to create the SSTable format easily.

The Cassandra record writer uses Cassandra's bulk-loader-like format to upload the output data back to Cassandra. Please make sure to configure the right settings as mentioned in the following sections to configure the right column family's in the MapReduce configuration to read and upload the data back to the right column families.

You can use the following configuration to change the MapReduce jobs. Most of the settings are self-explanatory and comments are also included for assistance.

A certain consistency level should be used for reads and writes of Thrift calls. You can choose to use Quorum if you want to make sure the data read and writes have strong consistency (repaired in most cases) before they are used for analytics. The default consistency level is `ConsistencyLevel.ONE`. The following settings allow you to configure the consistency levels:

- `cassandra.consistencylevel.read`
- `cassandra.consistencylevel.write`

Keyspace and column family settings

In addition to the previous settings, it might be useful to configure the following parameters for the MapReduce jobs to use (they are self-explanatory):

- `cassandra.input.partitioner.class` (defaults to `Murmur3Partitioner`)
- `cassandra.output.partitioner.class` (defaults to `Murmur3Partitioner`)
- `cassandra.input.keyspace`

- `cassandra.output.keyspace`
- `cassandra.input.keyspace.username`
- `cassandra.input.keyspace.passwd`
- `cassandra.output.keyspace.username`
- `cassandra.output.keyspace.passwd`
- `cassandra.input.columnfamily`

The following are MapReduce optimization parameters to read the data without causing memory pressure on the nodes:

- `cassandra.input.predicate`
- `cassandra.input.keyRange`
- `cassandra.input.split.size` (defaults to `64 * 1024`)

The following is a Boolean update if it is a wide row or not:

- `cassandra.input.widerows`

You can also change the Hadoop configuration to set a larger batch size. Using the following configuration reduces the network calls across Cassandra:

- `cassandra.range.batch.size` (defaults to `4096`)

Change the key size to be fetched for every read, which defaults to `8192`:

- `cassandra.hadoop.max_key_size`

Communication configuration using the Thrift interface with Cassandra

To configure the communication configurations for the MapReduce jobs to talk to Cassandra, you can change the following:

- `cassandra.input.thrift.port`
- `cassandra.output.thrift.port`
- `cassandra.input.thrift.address`
- `cassandra.output.thrift.address`
- `cassandra.input.transport.factory.class`
- `cassandra.output.transport.factory.class`
- `cassandra.thrift.framed.size_mb`
- `cassandra.thrift.message.max_size_mb`

Set the compression setting for the SSTable configuration so it can be streamed back to Cassandra:

- `cassandra.output.compression.class`
- `cassandra.output.compression.length`

HDFS location of the temporary files

The location of the files can be configured using the following:

- `mapreduce.output.bulkoutputformat.localdir`
- `mapreduce.output.bulkoutputformat.localdir`
- `mapreduce.output.bulkoutputformat.buffersize`

The throttle is on the amount of data that will be streamed to Cassandra. Streaming more means more overhead on the receiving end to compact the data and also to saturate the network:

- `mapreduce.output.bulkoutputformat.streamthrottlembits`

The maximum number of hosts that are allowed can be down during the BulkRecordWriter upload of SSTables:

- `mapreduce.output.bulkoutputformat.maxfailedhosts`

If you are noticing a lot of timeouts in Cassandra due to batch updates or you are reading the data out of Cassandra, it may be a good time to increase `rpc_timeout_in_ms` in `Cassandra.yaml`, or you can also change the batch sizes to reduce the amount of data that needs to be read or scanned. Since Cassandra doesn't support streaming, it is important to paginate the wide rows and also to be careful in reading the data out of the data store. For example, if you read 2 GB of data from a row and it is not paginated, it will naturally run out of memory since the data store has to first pull the data into memory and send it in one shot.

Summary

In this chapter we covered multiple ways to run Hadoop and other types of analytics jobs on top of the data stored on Cassandra. The key thing to take away from this chapter is understanding the importance of separating the analytics workload for better workload segregation. As a Layman tool, users can choose to extract all the data out of Cassandra using range queries, but this may not be optimal.

In the following chapter we will cover how Cassandra can be configured to secure the data stored in it.

7
Security and Troubleshooting

Security is a difficult topic; if handled haphazardly, it is as good as not doing it. In order to secure the data, users need to think about security in all levels. The first part is to encrypt the data, and secure it when it crosses the server boundaries. The second part is restricting access to unauthorized users.

Encryption

There are two levels of encryption: the first being encrypting data in flight and the second being encrypting data at rest. In some cases, you might want to encrypt both, but for most cases, you just need to encrypt the data across the firewall. Cassandra, out of the box, has a flexible encryption configuration. Cassandra supports a simple SSL/TSL socket level encryption from client-to-server and server-to-server encryption; enabling them is as easy as enabling them in the configuration file.

Users can enable server-to-server encryption by editing `Cassandra.yaml` and then enabling the `internode_encryption` file. Most use cases will require you to enable the encryption level to DC but not all, since the inter-DC traffic will be protected by a firewall. In order to create the certificates that will be used by Cassandra for internode encryption, perform the following steps.

Creating a keystore

The first step in enabling the encryption of traffic is to create a keystore to store the certificates in a secure location. To create the keystore, follow the given instructions:

1. Download your Organization Certificate / Certificate Chain / Generate one (`cassandra-app.cert`).

2. Log in to any Linux machine and run the following to create `p12`:

   ```
   $openssl pkcs12 -export -in cassandra-app.cert -inkey cassandra-app.key -certfile cassandra-app.cert -name "cassandra-app" -out cassandra-app.p12
   ```

3. Create the keystore (you might need the password at this stage):

   ```
   $ keytool -importkeystore -srckeystore cassandra-app.p12 -srcstoretype pkcs12 -destkeystore cassandra-app.jks -deststoretype JKS
   ```

4. List to make sure you have the right one:

   ```
   $ keytool -list -v  -keystore cassandra-app.jks -storepass <Password>
   ```

> Just in case you want to use a self-signed certificate, perform the following (before step 1):
>
> ```
> $openssl req -new -x509 -days 3650 -extensions v3_ca -keyout cassandra-app.key -out cassandra-app.cert -new
> ```
> ```
> $cp cassandra-app.key cassandra-app.key.org
> ```
> ```
> $openssl rsa -in cassandra-app.key.org -out cassandra-app.key
> ```

Creating a truststore

Perform all the steps as mentioned previously and you will have a truststore (name it sensibly to differentiate in the future). Run the following command to set a different password for the truststore:

```
$ keytool -import -keystore cassandra-app.truststore -file cassandra-app.cert -alias cassandra-app -storepass <diffrent pass>
server_encryption_options:
    internode_encryption: dc
    keystore: conf/cassandra-app.jks
    keystore_password: cassandra
```

```
truststore: conf/cassandra-app.truststore
truststore_password: cassandra
```

Alternatively, run the following command:

```
client_encryption_options:
    enabled: false
    keystore: conf/cassandra-app.jks
    keystore_password: cassandra
```

Transparent data encryption

There are few products that provide TDE; one such vendor is Gazzang. Gazzang helps organizations protect sensitive information in Apache Cassandra by transparently encrypting data in real time and providing advanced key management, which ensures only authorized processes access the data. Encryption is half the battle, and the other half of the battle is managing the key that is used to encrypt the file. For example, a decent disk level encryption will ensure that the file cannot be read as plain text, but reveals the key. So, if a hacker gets hold of the key, he/she can decrypt the files; hence, a solution like Gazzang can manage the key outside of the system and your data will be safer. Please note that most use cases will not require you to encrypt the data, unless the data store contains PCI information in it; in such a case, you can take a slight performance hit to secure the data. Encryption is not cheap, but it is required in some cases. The documentation on how to set it up can be found online.

Keyspace authentication (simple authenticator)

Enabling authorization requires configuration changes in Cassandra.yaml. Update authenticator from AllowAllAuthenticator to PasswordAuthenticator. The default super username is Cassandra and the password is cassandra.

The syntax is as follows:

```
CREATE USER user_name
  WITH PASSWORD 'password' NOSUPERUSER | SUPERUSER
GRANT permission_name PERMISSION
| GRANT ALL PERMISSIONS
   ON resource TO user
```

The illustration of the preceding syntax is as follows:

```
bin/cqlsh -u cassandra -p cassandra -k system
Connected to Athena_cluster_test at localhost:9160.
[cqlsh 3.0.2 | Cassandra 2.0-SNAPSHOT | CQL spec 3.0.0 | Thrift protocol
19.37.0]
Use HELP for help.
cqlsh:system> create user joe with password 'test';
cqlsh:Keyspace1> GRANT ALL ON KEYSPACE "Keyspace1" TO joe;
```

Once the cluster is set up and before the password authentication is enabled, make sure the system_auth keyspace is replicated. By default, it is configured as SimpleStategy with a replication factor of 1. By now, you should know what is wrong with it.

```
ALTER KEYSPACE system_auth WITH REPLICATION =
    {'class' : 'NetworkTopologyStrategy',  'DC1' : 3};
```

It is worth noting that the Cassandra keyspace authentication happens when a connection is established — as with any other database system — and hence it has a very low overhead; it is advisable to use it if needed.

In addition to the default configuration, if you would like to have a more powerful and centralized authentication mechanism, DSE packages come with Kerberos and LDAP authentication built in. Kerberos is a computer network authentication protocol that allows nodes to communicate over a non-secure network to prove their identity to one another in a secure manner using tickets. To understand how to configure the keyspace, please look into the DSE documentation.

JMX authentication

Sometimes, you might want to disable JMX access to the users, since JMX can be used to decommission, repair, and remove the node from the ring. If your organization requires that you protect the JMX with a username and password, you can do so using the following settings. Edit cassandra-env.sh to enable authentication:

```
JVM_OPTS="$JVM_OPTS -Dcom.sun.management.jmxremote.authenticate=true"
JVM_OPTS="$JVM_OPTS -Dcom.sun.management.jmxremote.password.
file=jmxremote.password"
```

Audit

Auditing DDL is done by default, where it logs the command or changes that are executed on the cluster to the logfile. Logging DML is really expensive and not recommended; by default, Cassandra doesn't log any of it, but if you do want to enable logging, enable the following in `log4j-server.properties`:

```
#log4j.logger.org.apache.cassandra.service.StorageProxy=DEBUG
```

It is easier to implement `IAuthenticator` than to add application logging if you would like to have a much better log of the information when the keyspace or column family authentication occurs and before any of the operations are executed on the keyspace or column family.

Things to look out for

One of the basic things to look for in the logs is the ERROR messages; if you notice any abnormalities in the logs, they should be addressed and understood. There are other things printed in the logs that have to be addressed as well. For example, as mentioned in the earlier chapter, Cassandra prints a log message if the garbage collection takes more than 200 ms. If you notice this, it is the right time to look at the historical GC metrics and understand whether we can tune to get a better GC performance. Users can also choose to enable debug on the GCInspector class to watch for all the pause times:

```
INFO [ScheduledTasks:1] 2013-05-19 15:50:08,708 GCInspector.java (line
116) GC for ConcurrentMarkSweep: 491 ms for 2 collections, 878049656
used; max is 1052770304
```

Cassandra also logs when the repair has started or is in progress. If you notice a repair hanging for a long time, you might want to correlate the events to see where it might be stuck. Most of the time, compactionstats and netstats will reveal more information. In addition to that, Cassandra logs information about the compaction and flush information that will be useful in understanding the load on the system. As an example, the following log message shows that the compaction has just started:

```
INFO [FlushWriter:1] 2013-05-19 15:56:37,258 Memtable.java (line 479)
Writing Memtable-Standard1@495929672(57111120/46137344 serialized/live
bytes, 1297980 ops)
INFO [CompactionExecutor:9] 2013-05-19 15:57:59,742 CompactionTask.
java (line 114) Compacting [SSTableReader(path='/var/lib/cassandra/
data/Keyspace1/Standard1/Keyspace1-Standard1-ja-4-Data.db'),
SSTableReader(path='/var/lib/cassandra/data/Keyspace1/Standard1/
Keyspace1-Standard1-ja-1-Data.db'), SSTableReader(path='/var/lib/
cassandra/data/Keyspace1/Standard1/Keyspace1-Standard1-ja-3-Data.
db'), SSTableReader(path='/var/lib/cassandra/data/Keyspace1/Standard1/
Keyspace1-Standard1-ja-2-Data.db')]
```

If you notice nodes flipping up and down all the time while executing `nodetool ring`, it means the node is overloaded. Therefore, `FailureDetector` marks the node to be down often enough causing issues for readers and writers. This will also mean that there will be additional latency seen by the client and requests can also fail. The following log shows that scenario:

```
INFO [GossipStage:1] 2013-03-26 10:34:17,024 Gossiper.java (line 786)
Node /17.209.115.136 has restarted, now UP
INFO [GossipStage:1] 2013-03-26 10:34:17,025 Gossiper.java (line 754)
InetAddress /17.209.115.136 is now UP
INFO [GossipStage:1] 2013-03-26 10:34:35,092 Gossiper.java (line 786)
Node /17.209.115.135 has restarted, now UP
```

Since Cassandra 2.0, there is no more two-phase compaction. If you are using an older version and notice that information, it is a good time to start optimizing your wide rows as they are too big. Cassandra logs when the node starts to drop hints, since we have past the hints window.

Summary

It is important to keep an eye on the direct memory usage. Since Cassandra 2.0, its internals relay more on off-heap data structures (bloom filters and index summaries). Having data in Cassandra doesn't imply zero overhead. It has to maintain the in-memory data structure for faster access. Having said that, you should also know that there are few tickets open to remove these data structures when the SSTables are not hot.

Traditional upgrading of RDBMS requires downtime since Cassandra 1.0. Upgrades for Cassandra do not need downtime and can support a cluster with multiple versions running on it. This required Cassandra to add versioning everywhere including the filesystem and messaging service. Hence, after the upgrade, the user needs to run `nodetool upgradesstables/scrub` to upgrade the SSTables. The limitation of this is we cannot bootstrap or repair nodes until we have a consistent version across the cluster. To be safer during the upgrade, the user might want to create a snapshot of the node before upgrading the node, and if something wrong happens, the user might be able to downgrade to the older version. The following is the help from the `nodetool` command:

```
 scrub [keyspace] [cfnames] - Scrub (rebuild sstables for) one or more
column families

 upgradesstables [-a|--include-all-sstables] [keyspace] [cfnames] -
Rewrite sstables (for the requested column families) that are not on the
current version (thus upgrading them to said current version).
```

When a network partition happens, Cassandra will try to store hints, so it can be replayed. This works fine for a blip, but if there is a prolonged network outage, it is better to run repairs to make sure the data is consistent across the datacenters. Using each_quorum is fine in most cases, until there is a network outage and the client nodes randomly drop. This can be addressed by adding some complicated logic in the client side to downgrade the consistency level to local_quorum.

It is true that most of the developed features first appear in OpenJDK before they appear on SunJDK, but in the past, the community has reported that OpenJDK has been really unstable.

The commit log is better served in a separate hard drive or on a separate device as mentioned in the earlier chapters. Since the I/O patterns are different, it is better to separate them. Using bigger batches (mutations and multigets) is not recommended, because the proxy will be busy coordinating a lot of traffic across the nodes. It has to wait for all of the operations to complete (including the slowest of the query) before responding back to the client, instead of parallelizing it. For example, instead of sending 1000 mutations in one batch, they can be sent by a 10-batch mutation parallelly, with a batch size of 100.

Sometimes, it will make sense to use **OPP (order preservative partitioner) / BOP (byte ordered partitioner)**. Try not to use these as there is a high possibility that they will create hotspots. Since the advent of vnodes, they do not have a lot of operational overheads, but you will notice a different performance characteristic. Avoid NFS or EBS volumes. The network latency required to complete disk seeks is really high.

Make sure all the keyspace with replication has NetworkTopologyStategy; this will save you from all the hassle of moving the nodes and data in the future (else you might need to decommission and bootstrap a node if you want to change it later on). Also, make sure to configure the right rack names and DC names (snitches) on the nodes, and make sure that the cluster is well balanced to begin with.

Cassandra doesn't need a load balancer, but if you do have one, make sure the load balancer is configured correctly for thrift connections. Most Cassandra drivers/clients come with round-robin and random node selection algorithms that will take care of load balancing. Make sure to choose the right one that fits your application.

Finally, the Linux configurations we talked about previously are more important for Cassandra to optimally utilize the server resources. In addition, there are more configurations such as large pages that will make some difference on some installations, but it requires a lot of testing as it differs with the hardware.

Index

Symbols

-Dreplace_tokens=<tokens> command 52

A

Acunu Analytics 85
ALTER TABLE 38
Apache Cassandra database 5
Audit 93
auto_bootstrap setting 15

B

backup configurations
  about 27
  auto_snapshot option 28
  incremental_backups option 28
backups analytics
  about 85
  column family settings 86, 87
  communication configuration 87
  file streaming 86
  Keyspace settings 86, 87
  temporary files HDFS location 88
basic JMX monitoring
  about 57
  JConsole and VisualVM 58
  Mx4J 58
  Push vs. Pull 58
BigTable data model
  about 6
  compactions 8
  Keyspace 7
  Memtable 7, 8
  SSTable 7
binaries
  downloading, for installation 14
  selecting, for installation 14
bloom filter 12
bootstrapping
  about 52, 53
  vnodes 53
broadcast_address setting 16
ByteOrderedPartitioner 15

C

CAP theorem
  about 5
  availability 5
  consistency 5
  partition tolerance 5
Cassandra
  about 5, 11
  data consistent features 49
  data, reading, directly from 85
  eventual consistency 10
  feature 9
  Hadoop, configuring with 81
  in-memory data structure 12
  memory requisites 12
  monitoring tools 56
  network requisites 12
  on EC2 instance 28, 29
  performance affecting, factors 61
Cassandra data models
  counter column family, creating 46
  denormalization 43
  secondary index 46
  tweet data structure 46
cassandra-env.sh
  configuring 14

Cassandra multi region cluster
  configuring 78-80
Cassandra Query Language (CQL) 37
Cassandra.yaml file
  configuring 15
Cfhistograms 54
cfstats command 57
cleanup 54
cluster_name setting 15
column family
  about 6
  compaction 33
  configuration options 36
  creating 31, 32
  GC grace period 32
  maximum compaction threshold 33
  minimum compaction threshold 33
column_index_size_in_kb setting 20
column, types
  composite columns 42
  counter columns 42
  expiring columns 42
  standard columns 42
commit log
  separate drive 75
commitlog_directory setting 16
commitlog_segment_size_in_mb setting 18
commitlog_sync_batch_window_in_ms
    method 18
commitlog_sync_period_in_ms method 18
commitlog_sync setting
  about 18
  commitlog_sync_batch_window_in_ms
    method 18
  commitlog_sync_period_in_ms method 18
commitlog_total_space_in_mb setting 18
compaction_preheat_key_cache setting 19
compactions
  about 8
  LeveledCompactionStrategy 70, 71
  SizeTieredCompactionStrategy 70
compactionstats command 57
compaction_throughput_mb_per_sec
    setting 20
composite columns 42
composite primary key type 34, 35
compression

Cassandra cache 73, 74
Cassandra kernel tuning 75
Cassandra multi region cluster, configuring
    78-80
dynamic snitch 77, 78
filesystem caches 75
Increase Map Count parameter 77
Increase User limits parameter 77
netstats 72
NOOP scheduler 76
NUMA 76
tpstats 72, 73
compression metadata 12
concurrent_compactors setting 21
ConcurrentLinkedHashCacheProvider 20
concurrent_reads setting 21
concurrent_writes setting 22
configuration, cassandra-env.sh 14
configuration, Cassandra.yaml file
  about 15
  auto_bootstrap setting 15
  broadcast_address setting 16
  cluster_name setting 15
  column_index_size_in_kb setting 20
  commitlog_directory setting 16
  commitlog_segment_size_in_mb setting 18
  commitlog_sync setting 18
  commitlog_total_space_in_mb setting 18
  compaction_preheat_key_cache setting 19
  compaction_throughput_mb_per_sec
    setting 20
  concurrent_compactors setting 21
  concurrent_reads setting 21
  concurrent_writes setting 22
  data_file_directories setting 16
  disk_failure_policy setting 16
  endpoint_snitch setting 18
  flush_largest_memtables_at setting 22
  index_interval setting 22
  initial_token setting 17
  in_memory_compaction_limit_in_mb
    setting 21
  key cache, saving to disk 19
  key_cache_size_in_mb setting 19
  listen_address setting 17
  memtable_flush_queue_size setting 23
  memtable_flush_writers setting 23

memtable_total_space_in_mb setting 22
partitioner setting 15
populate_io_cache_on_flush setting 21
ports settings 17
request_scheduler_options setting 23
request_scheduler setting 23
row_cache_provider setting 19
row cache, saving to disk 19
row_cache_size_in_mb setting 19
rpc_address setting 17
rpc_keepalive setting 24
rpc_max_threads setting 24
rpc_min_threads setting 25
rpc_server_type setting 24
seed_provider setting 15
stream_throughput_outbound_megabits_
    per_sec setting 23
thrift_framed_transport_size_in_mb
    setting 24
timeouts 25
configuration options, column family
chunk_length_kb 36
crc_check_chance 36
sstable_compression 36
counter column 42
counter column family
creating 46
CREATE TABLE 38

D

data
querying 37
data consistent features, Cassandra
Hinted handoff 49
manual repair 50
data_file_directories setting 16
data modeling
Dynamic columns 41
fixed columns 41
data querying
about 37
ALTER TABLE 38
CREATE TABLE 38
DESCRIBE command 39
SELECT 39
USE statement 37

DataStax
reference documentation 37
URL 14
DataStax OpsCenter 56
dclocal_read_repair_chance parameter 36
decommission 54
denormalization
about 43
comparator 44
distributed counters example 45
key-value data store example 45
Time series database example 44
DESCRIBE CLUSTER option 39
DESCRIBE command 39
DESCRIBE SCHEMA option 39
disk_failure_policy setting 16
distributed Hash table 9
domain-specific language (DSL) 37
drain 54
DSE Hadoop 84
dynamic snitch
about 26, 77, 78
dynamic_snitch_badness_threshold
    parameter 26
dynamic_snitch_reset_interval_in_ms
    parameter 26
dynamic_snitch_update_interval_in_ms
    parameter 26
hinted_handoff_enabled parameter 26
hinted_handoff_throttle_delay_in_ms
    parameter 27
inter_dc_tcp_nodelay parameter 27
max_hints_delivery_threads parameter 27
max_hint_window_in_ms parameter 27
phi_convict_threshold parameter 26
dynamic_snitch_badness_threshold
    parameter 26
dynamic_snitch_reset_interval_in_ms
    parameter 26
dynamic_snitch_update_interval_in_ms
    parameter 26

E

Ec2MultiRegionSnitch 29
Ec2Snitch 29
encryption

keystore, creating 90
levels 89
transparent data encryption 91
truststore, creating 90
endpoint_snitch setting 18
eventual consistency 10
expiring columns 42

F

file streaming 86
flush_largest_memtables_at setting 22

G

ganglia 58
GCInspector class 93
GCLogs
jmap tool 68
jstat tool 66, 67
Stop-the-world GC 66
GC logging
creating 65
GossipingPropertyFileSnitch 83
Gossip protocol 8

H

Hadoop
configuring, with Cassandra 81
Hadoop, configuring with Cassandra
about 82
virtual datacenter, configuring 82
Hinted handoff
about 49
down side 49
hinted_handoff_enabled parameter 26
hinted_handoff_throttle_delay_in_ms
parameter 27

I

Increase Map Count parameter 77
Increase User limits parameter 77
index_interval setting 22
index summary 12
initial_token setting 17
in_memory_compaction_limit_in_mb

setting 21
in-memory data structure, Cassandra
bloom filter 12
compression metadata 12, 13
index summary 12
key cache 13
row cache 14
SSDs versus spinning disks 13
inter_dc_tcp_nodelay parameter 27
iostat 62, 63

J

JConsole and VisualVM 58
jmap
URL 13
JMX authentication 92

K

key cache 13
keyspace
column family, creating 31, 32
creating 30
keystore
creating 90

L

listen_address setting 17
logs
errors, checking 93

M

manual repair
disadvantage 51
streaming 51
validation compaction stage 50
max_hints_delivery_threads parameter 27
max_hint_window_in_ms parameter 27
Mean Time Between Failures (MTBF) 5
Memtable 7, 8
memtable_flush_queue_size setting 23
memtable_flush_writers setting 23
memtable_total_space_in_mb setting 22
Merkle tree 50
monitoring tools, Cassandra

basic JMX monitoring 57, 58
DataStax OpsCenter 56
move <token> 54
Murmur3Partitioner random partitioner 15
Mx4J 58

N

netstats 57, 72
netstats <host> 55
NetworkToplogyStrategy (NTS) 82
nodetool command 27, 56
nodetool --help command 54
noop scheduler 76
NUMA 76

P

partitioner
  about 15
  ByteOrderedPartitioner 15
  Murmur3Partitioner random partitioner 15
  RandomPartitioner 15
phi_convict_threshold parameter 26
populate_io_cache_on_flush setting 21
ports
  about 17
  native_transport_port option 17
  rpc_port option 17
  storage_port option 17
PropertyFileSnitch 83

Q

query performance
  tracing 40

R

RandomPartitioner 15
range_request_timeout_in_ms setting 25
read operations, drain
  move <token> 54
  netstats <host> 55
  rebuild <dc_name> 55
  removenode 55
  scrub 55
  setcompactionthroughput 55

setstreamthroughput 55
snapshot 55
stop 56
Tpstats 56
read_repair_chance parameter 36
read_request_timeout_in_ms setting 25
rebuild <dc_name> 55
removenode 55
request_scheduler method
  about 23
  org.apache.cassandra.scheduler.
    NoScheduler 23
  org.apache.cassandra.scheduler.
    RoundRobinScheduler 23
request_scheduler_options method
  about 23
  default_weight 23
  throttle_limit 23
request_timeout_in_ms setting 25
row cache 14
row_cache_provider setting 19
rpc_address setting 17
rpc_keepalive setting 24
rpc_max_threads setting 24
rpc_min_threads setting 25
rpc_server_type setting
  about 24
  hsha option 24
  sync option 24

S

scrub 55
secondary index
  creating 47
  gotchas 46
  indexed column 47
  internal data structure 47
  table, creating 47
secondary indexes 34, 38
security 89
seed_provider setting 15
SELECT 39
SerializingCacheProvider 19
setcompactionthroughput 55
snapshot 55
snitches, Cassandra

Ec2MultiRegionSnitch 29
Ec2Snitch 29
Sorted String Table. *See* **SST table**
spinning disks
   versus SSDs 13
SSDs
   versus spinning disks 13
SSTable 7
standard columns 42
Stop 56
streaming_socket_timeout_in_ms setting 25
stream_throughput_outbound_megabits_
        per_sec setting 23

T

thrift_framed_transport_size_in_mb setting
        24
timeouts 25
Timestamp 6
tpstats 56, 57, 72, 73
transparent data encryption
   about 91
   JMX authentication 92
   keyspace authentication 91, 92
truncate_request_timeout_in_ms setting 25
truststore
   creating 90

U

USE statement 37

V

virtual datacenter configuration
   about 82
   DSE Hadoop 84
   GossipingPropertyFileSnitch 83
   PropertyFileSnitch 82, 83
Visual VM
   URL 13
vmstat 62

vnodes
   about 53
   cfhistograms 54
   cleanup 54
   decommission 54
   drain 54
   node tool commands 54

W

write_request_timeout_in_ms setting 25

Thank you for buying
Learning Cassandra for Administrators

About Packt Publishing

Packt, pronounced 'packed', published its first book "*Mastering phpMyAdmin for Effective MySQL Management*" in April 2004 and subsequently continued to specialize in publishing highly focused books on specific technologies and solutions.

Our books and publications share the experiences of your fellow IT professionals in adapting and customizing today's systems, applications, and frameworks. Our solution based books give you the knowledge and power to customize the software and technologies you're using to get the job done. Packt books are more specific and less general than the IT books you have seen in the past. Our unique business model allows us to bring you more focused information, giving you more of what you need to know, and less of what you don't.

Packt is a modern, yet unique publishing company, which focuses on producing quality, cutting-edge books for communities of developers, administrators, and newbies alike. For more information, please visit our website: www.packtpub.com.

About Packt Open Source

In 2010, Packt launched two new brands, Packt Open Source and Packt Enterprise, in order to continue its focus on specialization. This book is part of the Packt Open Source brand, home to books published on software built around Open Source licences, and offering information to anybody from advanced developers to budding web designers. The Open Source brand also runs Packt's Open Source Royalty Scheme, by which Packt gives a royalty to each Open Source project about whose software a book is sold.

Writing for Packt

We welcome all inquiries from people who are interested in authoring. Book proposals should be sent to author@packtpub.com. If your book idea is still at an early stage and you would like to discuss it first before writing a formal book proposal, contact us; one of our commissioning editors will get in touch with you.

We're not just looking for published authors; if you have strong technical skills but no writing experience, our experienced editors can help you develop a writing career, or simply get some additional reward for your expertise.

[PACKT] open source ✤
PUBLISHING
community experience distilled

Cassandra High Performance Cookbook

ISBN: 978-1-84951-512-2   Paperback: 310 pages

Over 150 recipes to design and optimize large-scale Apache Cassandra deployments

1. Get the best out of Cassandra using this efficient recipe bank

2. Configure and tune Cassandra components to enhance performance

3. Deploy Cassandra in various environments and monitor its performance

Mastering Apache Cassandra

ISBN: 978-1-78216-268-1   Paperback: 340 pages

Get comfortable with the fastest NoSQL database, its architecture, key programming patterns, infrastructure management, and more!

1. Complete coverage of all aspects of Cassandra

2. Discusses prominent patterns, pros and cons, and use cases

3. Contains briefs on integration with other software

Please check **www.PacktPub.com** for information on our titles

[PACKT] open source ✳
PUBLISHING   community experience distilled

Instant Apache Cassandra for Developers Starter

ISBN: 978-1-78216-390-9   Paperback: 50 pages

Start develpoing with Cassandra and learn how to handle big data

1. Learn something new in an Instant! A short, fast, focused guide delivering immediate results

2. Tune and optimize Cassandra to handle big data

3. Learn all about the Cassandra Query Language and Cassandra-CLI

Instant Cassandra Query Language

ISBN: 978-1-78328-271-5   Paperback: 54 pages

A Partical, step-by-step guide for quickly getting started with Cassandra Query Language

1. Learn something new in an Instant! A short, fast, focused guide delivering immediate results

2. Covers the most frequently used constructs using practical examples

3. Dive deeper into CQL, TTL, batch operations, and more

Please check **www.PacktPub.com** for information on our titles